ITERATED INTEGRALS
AND CYCLES ON
ALGEBRAIC MANIFOLDS

NANKAI TRACTS IN MATHEMATICS

Series Editors: Shiing-shen Chern, Yiming Long, and Weiping Zhang
Nankai Institute of Mathematics

Published

Nankai Tracts in Mathematics – Vol. 7

ITERATED INTEGRALS AND CYCLES ON ALGEBRAIC MANIFOLDS

Bruno Harris

Department of Mathematics
Brown University, USA

World Scientific
New Jersey • London • Singapore • Hong Kong

Published by

World Scientific Publishing Co. Pte. Ltd.

5 Toh Tuck Link, Singapore 596224

USA office: Suite 202, 1060 Main Street, River Edge, NJ 07661

UK office: 57 Shelton Street, Covent Garden, London WC2H 9HE

British Library Cataloguing-in-Publication Data
A catalogue record for this book is available from the British Library.

ITERATED INTEGRALS AND CYCLES ON ALGEBRAIC MANIFOLDS
Nankai Tracts in Mathematics — Vol. 7

Copyright © 2004 by World Scientific Publishing Co. Pte. Ltd.

ISBN 981-238-720-X

Printed in Singapore by World Scientific Printers (S) Pte Ltd

To Professor S.S. Chern and

to the memory of Professor K.T. Chen

Preface

This book begins with some work of K.T.Chen, who used iterated integrals of 1-forms on a manifold to study its fundamental group π_1, more precisely the quotients of π_1 by the sequence of groups in its lower central series: $\pi_1/(\pi_1, \pi_1)$, $\pi_1/((\pi_1, \pi_1), \pi_1)$, and so on, $(,)$ denoting the commutator subgroup. The first step is well-known: if we choose a vector space (over $\mathbb{R}$) consisting of closed 1-forms and isomorphic to the first deRham cohomology group of the manifold X then integration of these forms over closed paths gives a homomorphism of $\pi_1(X)$ to the $\mathbb{R}$-dual of $H^1(X)$, which is just $H_1(X, \mathbb{R})$ if we assume H_1 finite dimensional, and with this assumption $\pi_1/(\pi_1, \pi_1)$ (mod torsion) embeds as a discrete cocompact subgroup in the Lie group $H_1(X, \mathbb{R})$. Chen showed that if one uses iterated integrals of the closed 1-forms, e.g integrals of the form

$$\int_{t_2=0}^{1} [\int_{t_1=0}^{t_2} f_1(t_1) dt_1] f_2(t_2) dt_2$$

in the case of two 1-forms, then one can similarly map $\pi_1/((\pi_1, \pi_1), \ldots, \pi_1)$ to a nilpotent Lie group (constructed using $H_1(X)$ and $H_2(X)$) and obtain information on this quotient of π_1; again the image is discrete and cocompact.

We will explain Chen's results in the most concrete form for a compact Kahler manifold X, where the whole sequence of π_1 quotients and their homomorphism to Lie groups is obtained by means of a flat connection θ canonically associated to X. θ is a 1-form on X with values in an (infinite-dimensional) Lie algebra (or more precisely a formal series of 1-forms on X with Lie algebra coefficients). This is done in Chapter 1, and Chapter 2 studies more closely the special case where X is a compact Riemann surface and only the part of the above construction involving iterated integrals of two 1-forms is examined. It turns out that these iterated integrals give information about how the Riemann surface X is embedded in its Jacobian $J(X)$. Following A.Weil [Weil 1962, p.331] we consider both X and its image X^- under the map of $J(X)$ to itself which is the group-theoretic inverse, and form the algebraic 1-cycle $X - X^-$, which is homologous to 0 in $J(X)$. This 1-cycle, following Hodge and Weil, has an image in another torus, an intermediate Jacobian of $J(X)$ associated to $H^3(J(X))$. Weil puts this example in a discussion of whether this algebraic 1-cycle homologous to zero is also algebraically equivalent to 0 (roughly speaking, whether it can be deformed to 0 using an algebraic deformation). We show that just this "quadratic part" of Chen's construction calculates (gives a formula for) the "Abel-Jacobi" image of $X - X^-$ in the intermediate Jacobian. This allows us to prove that "in general" $X - X^-$ is not algebraically equivalent to 0 and also to give specific examples, the first specific examples of cycles homologous but not algebraically equivalent to 0, consisting of algebraic curves over $\mathbb{Z}$ such as the Fermat curve $x^4 + y^4 = 1$. We refer to our papers [Harris 1983a], [Harris 1983b].

Chapter 3 (partially) generalizes Chapter 2 to higher dimensional Kahler manifolds X. In the Riemann surface case the construction in Chapter 2 associates to three elements of $H_1(X, \mathbb{Z})$ represented by mutually disjoint 1-cycles C_1, C_2, C_3 a real number obtained by iterated integrals over one of them, say C_3, of harmonic forms α_1, α_2 Poincare dual to C_1, C_2. If we take twice this real number and reduce it mod $\mathbb{Z}$, and do this for all possible triples of homology classes as above, we can regard this set of real numbers mod $\mathbb{Z}$ as the Abel-Jacobi image of $X - X^-$. The generalization in Chapter 3 consists in taking k homology classes $[C_1], \ldots, [C_k]$ (of possibly different dimensions) representable by cycles $C_1, \ldots, C_k$ such that any $k-1$ of the C_i are disjoint (so $k \geq 3$) and

$$\sum_{i=1}^{k} codim C_i = (dim X) + 1$$

and associating to this k-tuple of homology classes a real number, which depends only on the complex structure of X (if X has a Kahler metric). This real number is constructed by using the heat kernel $\exp(-t\Delta)$ of the Laplace operator Δ on differential forms. We show that this same number can be expresses by iterated integrals involving the harmonic forms Poincare dual to the $[C_i]$, where the domains of integration are now intersections of the cycles C_i.

We have attempted throughout the book to give definitions and details of proofs that will make it accessible to, say, second year graduate students, or perhaps even students with less background. We do however assume some acquaintance with topology and with Lie groups in Chapter 1. In Chapter 3 we explain a connection between the heat kernel and cycle pairings on a Riemannian manifold and rely considerably on the second chapter of "Heat Kernels and Dirac Operators" by Berline, Getzler, and Vergne. For K.T.Chen's work we can refer to his collected works and the article by Richard Hain (see Bibliography).

I would like to express my gratitude to Professor Chern Shiing-Shen for inviting me to present these results at the Conference on the Work of K.T.Chen and W.L.Chow at the Nankai Mathematical Institute in October 2000, to come back in Fall 2001 to give a course on this subject, and to write the present book. Professor Chern's warm kindness and that of other faculty and students at Nankai during this visit have made it an unforgettable experience: I would like to mention especially Professors Zhou Xing-Wei, Bai Chengming, Feng Huitao, Long Yiming, Fang Fuquan, Ge Molin, and students Miss Long Jing and Miss Zhu Tong.

For a similar course at Brown in Fall 2002, I would like to thank Amir Jafari, Wang Qingxue, Justin Corvino, and Alan Landman. For some of the basic mathematical knowledge in this book, I am grateful to my friends H.C.Wang, Gerard Washnitzer, and Ezra Getzler.

Contents

Iterated Integrals, Chen's Flat Connection and π_1

1.1 Introduction

Iterated integrals in calculus have the form

$$I = \int_{t=0}^{x} \Big[\int_{u=0}^{t} f_1(u)du \Big] f_2(t)dt. \tag{1.1}$$

More generally

$$I = \int \cdots \int_{0 \le t_1 \le \cdots \le t_n \le x} f_1(t_1)dt_1 f_2(t_2)dt_2 \cdots f_n(t_n)dt_n. \tag{1.2}$$

Interesting examples: take

$$f_1(t)dt = \frac{dt}{1-t}, \quad f_2 dt = \cdots = f_r dt = \frac{dt}{t},$$

$$I = \sum_{n=1}^{\infty} \frac{x^n}{n^r},$$

$$I \to \sum \frac{1}{n^r} = \zeta(r) \quad \text{as } x \to 1.$$

Iterated integrals are related to differential equations, which in turn are related to π_1, the fundamental group of a manifold: this latter relation involves Lie groups and Lie algebras. So we will begin by proving a theorem of K.T. Chen which describes nilpotent quotients of π_1: $\pi_1/(\pi_1, \pi_1), \pi_1/((\pi_1, \pi_1), \pi_1), \cdots$ using certain nilpotent Lie groups. The rest of the book will examine more closely the iterated integrals which give this relationship: "Chen's flat connection" for π_1 of compact Kahler manifolds , or more specifically complex projective algebraic manifolds, and find that we obtain some information on algebraic cycles and on homology of these manifolds. We will introduce a little Hodge theory and the heat operator $e^{-t\Delta}$, Δ = Laplacian on forms. The more experienced reader should skip much of the introductory material on Lie theory.

1.2   Differential equations

On the real line $\mathbb{R}$, consider the $n \times n$ first order linear system

$$\frac{d\vec{y}}{dt} = \vec{y}(t)A(t) \quad (\text{initial condition } \vec{y}(0) = \vec{c}) \tag{1.3}$$

$\vec{y} = (y_1, \cdots, y_n)$, $A(t)$ = given $n \times n$ matrix of functions. If $A(t)$ is a constant matrix A then the solution is $\vec{y}(t) = \vec{c}e^{At}$. If we take $\vec{c}$ = standard basis vector $\vec{e}_i$, and $\vec{y}_i(t) = \vec{e}_i \exp(At)$ then we can put these n solution vectors into the n rows of an invertible matrix $Y(t)$ and rewrite the problem as: find $Y(t)$ invertible satisfying

$$\begin{aligned} dY(t) &= Y(t)A(t)dt \\ Y(0) &= I. \end{aligned} \tag{1.4}$$

The solution is given by an infinite series of iterated integrals

$$Y(t) = I + \int_0^t A(t_1)dt_1 + \int_0^t \left[\int_0^{t_2} A(t_1)dt_1 \right] A(t_2)dt_2$$
$$+ \cdots + \int \cdots \int_{0 \leq t_1 \leq \cdots \leq t_n \leq t} A(t_1)dt_1 A(t_2)dt_2 \cdots A(t_n)dt_n + \cdots$$

The series converges if $A(t)$ is continuous in t, and we can differentiate term by term to verify (1.4).

Next we take X to be any n-dimensional real differentiable manifold, $G = GL(N, \mathbb{R})$, or any other Lie group, and $\mathfrak{g}$ = Lie algebra of G, $\mathfrak{g} = M(N, \mathbb{R})$ (all $N \times N$ matrices). $\mathfrak{g}$ can be identified with the tangent space $T_e(G)$ at the identity e; given $C \in \mathfrak{g}$ we form $e^{tC} = I + tC + \cdots$, which gives us a path in G (a 1-dimensional Lie group) and C is its tangent vector at e.

Next we define a *1-form α on X with values in $\mathfrak{g}$* to be a finite sum

$$\alpha = \sum_{i,j=1}^N \alpha_{ij} \otimes_{\mathbb{R}} C_{ij}$$

where α_{ij} are ordinary (scalar valued) 1-forms on X and C_{ij} are in $\mathfrak{g}$. We can in the case where $\mathfrak{g} = M(N, \mathbb{R})$, regard α as an $N \times N$ matrix with 1-form entries. We will write the matrix product of such matrices as $\alpha \wedge \beta$. Note that $\alpha \wedge \alpha$ need not be 0 if $N > 1$. If $\mathfrak{g}$ is a general Lie algebra we cannot multiply in this way as $\mathfrak{g}$ is only closed under the $[,]$ product. However we define, for $\alpha = \sum \alpha_i \otimes C_i$, $\beta = \sum \beta_j \otimes C'_j$ ($C_i, C'_j \in \mathfrak{g}, \alpha_i, \beta_j$ 1-forms $\in A^1(X)$), the bracket $[\alpha, \beta]$ as

$$[\alpha, \beta] = \sum_{i,j} \alpha_i \wedge \beta_j \otimes [C_i, C'_j].$$

If $\mathfrak{g} = M(N, \mathbb{R})$ we calculate that $\alpha \wedge \alpha = \frac{1}{2}[\alpha, \alpha]$. (So, $[\alpha, \alpha]$ need not be 0). We define $[\alpha, \beta](v, w) = [\alpha(v), \beta(w)] - [\alpha(w), \beta(v)]$. We can also define $d\alpha = \sum_i d\alpha_i \otimes C_i$ as a $\mathfrak{g}$-valued 2-forms. As a function of pairs v, w of vector fields,

$$d\alpha(v, w) = v(\alpha(w)) - w(\alpha(v)) - \alpha([v, w]).$$

Here $\alpha(v) = \sum \alpha(v_i)c_i$ is a $\mathfrak{g}$-valued function on X, and $w(\alpha(v))$ is the $\mathfrak{g}$-valued function obtained by acting with w (on the scalar function part of $\alpha(v)$). The main example is the *left-invariant* Maurer-Cartan 1-form μ on a Lie group G with Lie algebra $\mathfrak{g}$. Recall first that we can define $\mathfrak{g} = T_e(G)$ and also identify a tangent vector $v_e \in T_e(G)$ with the vector field $v_g = L_{g*}(v_e)$ (L_{g*}=action on tangent vectors of left translation by g, with L_g). v_g is then a left-invariant vector field: $L_{g'*}(v_g) = v_{g'g}$, and we have an isomorphism(of Lie algebras) $T_e(G) \to$ Lie algebra of all left invariant vector fields, by $v_e \mapsto L_g(v_e) = v_g$.

We now define the $T_e(G) = \mathfrak{g}$-valued 1-form μ on G: for $g \in G$, $v_g \in T_g(G)$, $\mu(v_g) = L_{g^{-1}*}(v_g)$. So if $v_g = L_{g*}(v_e)$ then $\mu(v_g) = v_e$.

Proposition 1.1 *The Maurer-Cartan 1-form μ satisfies*

$$d\mu + \frac{1}{2}[\mu, \mu] = 0 \quad \text{(as 2-form on } G\text{)}.$$

Proof: We first evaluate these 2-forms on pairs of *left-invariant* vector fields v, w. Note that $\mu(w)$ is a constnat (in g) $\mathfrak{g}$-valued functions on G, and so $v(\alpha(w)) = 0$. Thus

$$d\mu(v, w) = -\mu([v, w]) \in \mathfrak{g}.$$

Recall that we defined

$$[\mu, \mu](v, w) = [\mu(v), \mu(w)] - [\mu(w), \mu(v)] = 2[\mu(v), \mu(w)].$$

(the definition $[\alpha, \beta](v, w) = [\alpha(v), \beta(w)] - [\alpha(w), \beta(v)]$ is consistent with $\alpha = \sum \alpha_i \otimes C_i$, $\beta = \sum \beta_j \otimes C_j$ and $\alpha_i \wedge \beta_j (v, w) = \alpha_i(v) \beta_j(w) - \alpha_i(w) \beta_j(v)$).
Now $\mu(v) = v_e$ by the definition of μ for a left-invariant v, so

$$d\mu(v, w) = -\mu([v, w]) = -[v, w]_e = -[v_e, w_e],$$

and

$$\frac{1}{2}[\mu, \mu](v, w) = [\mu(v), \mu(w)] = [v_e, w_e].$$

Thus

$$(d\mu + \frac{1}{2}[\mu, \mu])(v, w) = 0$$

for left-invariant vector fields, but any vector field $\xi = \sum f_i v_i$, f_i functions and v_i left-invariant, and 2-forms are linear over (scalar-valued) functions. $\square$

Remark 1.1 *μ is frequently written as the 1-form $g^{-1}dg$ on G. This can be thought about as follows. g denotes the map $G \to G$ which takes g to g, i.e., the identity map dg is then the differential of the identity map, so for $v_g \in T_g$, $dg(v_g) = v_g$. $g^{-1}dg$ then stands for the composite $L_{g^{-1}*} \circ dg : v_g \mapsto L_{g^{-1}*}(v_g)$. Using this notation, $g^{-1}dg \wedge g^{-1}dg$ is a matrix product (if $G = GL(N, \mathbb{R})$) and*

$$d(g^{-1}dg) = d(g^{-1}) \wedge dg = -g^{-1}dg \wedge g^{-1}dg,$$

so $d(g^{-1}dg) + g^{-1}dg \wedge g^{-1}dg = 0$. We see now that for any map $Y : X \to G$, the pull-back $Y^*\mu = \alpha$ is a $\mathfrak{g}$-valued 1-form on X and satisfies $d\alpha + \frac{1}{2}[\alpha, \alpha] = 0$.

Theorem 1.1 *If X is a manifold, α a $\mathfrak{g}$-valued 1-form on X satisfying $d\alpha + \frac{1}{2}[\alpha, \alpha] = 0$, $x_0 \in X$ a given point, and G is a Lie group with Lie algebra $\mathfrak{g}$, then*

1. There exists an open neighborhood U of x_0 in X and a map $Y : U \to G$ such that

$$i)\ Y^*(\mu) = \alpha \quad and \quad ii)\ Y(x_0) = e.$$

2. If U is a connected neighborhood of x_0 in X and $Y : U \to G$ satisfies i) and ii) of 1., then Y is unique and is given by the following formula; if

$p : [0,1] \to U$ *is a path from* x_0 *to* x *in* U, *then if* $G = GL(N, \mathbb{R}) \subset M(N, \mathbb{R})$
or if G *is a Lie group in an associative algebra* A *with identity element* e,
then

$$Y(x) = e + \int_0^1 p^*\alpha(t_1) + \iint_{0 \le t_1 \le t_2 \le 1} p^*\alpha(t_1) \wedge p^*\alpha(t_2) + \cdots$$

(each term is in A, *sum is in* G). *If* p, p' *are homotopic paths from* x_0 *to* x
in U *(fixed end points during homotopy) then the iterated integral formulas
over* p *and over* p' *are equal.*

    3. If X *is connected and simply connected,* $x_0 \in X$, *then there is a
unique* $Y : X \to G$ *satisfying i) and ii) of 1. and given by 2.*

    If X *is not assumed simply-connected and* $\pi : \tilde{X} \to X$ *is its universal
covering with* $\tilde{x}_0 \in \tilde{X}$, $\pi(\tilde{x}_0) = x_0$ *given, then we let* $\tilde{\alpha}$ *on* $\tilde{X}$ *be* $\pi^*\alpha$. *Then
there exists* $\tilde{Y} : (\tilde{X}, \tilde{x}_0) \to (G, e)$ *as above and* $\tilde{Y}$ *gives a homomorphism
$m : \pi_1(X, x_0) \to G$ *such that* $\tilde{Y}$ *is equivariant. More generally,* $\tilde{Y}$ *gives a
homomorphism of the fundamental groupoid of* X *into* G.

    An element of the fundamental groupoid is a path $p : [0,1] \to X$ *with
$p(0) = x, p(1) = x'$, up to homotopy. The homomorphism sends* p *into an
element* $g \in G$ *as follows: there is a unique map* $Y_p : [0,1] \to G$ *such that
$Y_p^*(\mu) = p^*(\alpha)$ *on* $[0,1]$ *and* $Y_p(0) = e$; *we let* $g = Y_p(1)$. *If* $p : [0,1] \to Y$
and $p' : [0,1] \to Y$ *are composable paths, i.e.,* $p(1) = p'(0)$ *and* p *is sent to
g, p' *to* g' *then the composite path* pp' *is sent to* gg'.

Proof: We concentrate on proving the following statement: there is a
covering space $\pi : \tilde{X} \to X$ of X and a map $\tilde{Y} : \tilde{X} \to G$ such that $\tilde{Y}^*\mu = \pi^*\alpha$
and $\tilde{Y}(\tilde{x}_0) = e$ for some point $\tilde{x}_0 \in \tilde{X}$ above $x_0 \in X$. The other statements
are easy consequences of covering space theory, while the iterated integral
formula is just the 1-dimensional case and has already been discussed.

    The idea of the proof is to replace maps (defined locally) $X \to G$ by
their graphs in $X \times G$, and to replace the requirements on the maps by
specifications on the tangent spaces to these graphs. Existence of such
graphs will follow from the Frobenius integrability theorem.

    Thus on $X \times G$ we form the $\mathfrak{g}$-valued 1-form $A = pr_1^*(-\alpha) + pr_2^*(\mu)$
(where pr_i, $i = 1, 2$, are the projections to X, G). We write for short,
$A = -\alpha + \mu$. The tangent space to $X \times G$ at any point (x, g) is denoted
$T_xX \oplus T_gG$, with elements (v_x, w_g).

    Let $D_{x,g} \subset T_xX \oplus T_gG$ be the subspace of all (v_x, w_g) satisfying $\alpha(v_x) =
\mu(w_g)$, i.e., $A(v_x, w_g) = 0$. Since $\mu : T_g(G) \to \mathfrak{g}$ is an isomorphism, the
dimension of $D_{x,g}$ is the same as the dimension of T_xX, therefore constant.

In fact $D_{x,g}$ is the graph of a linear map $T_x X \to T_g G$ which pulls back μ_g to α_x.

To satisfy the hypotheses of the Frobenius integrability theorem, we have to show that if V_1, V_2 are vector fields on $X \times G$ which lie in $D_{x,g}$ at every (x, g), then $[V_1, V_2]$ also lies in $D_{x,g}$ at every (x, g). We can either do this by a direct calculation, or else prove an equivalent condition: the "matrix coefficients" a_i of the $\mathfrak{g}$-valued 1-form $A = -\alpha + \mu$, $A = \sum a_i \otimes C_i$ where the c_i are any basis of $\mathfrak{g}$, satisfy: each $da_i \in$ ideal (in the algebra of forms on $X \times G$) generated by $a_1, a_2, \cdots$; in other words $da_i = \sum a_j \wedge b_{ji}$ for some 1-forms b_{ji}.

To see this we write

$$dA = -d\alpha + d\mu = \frac{1}{2}([\alpha, \alpha] - [\mu, \mu])$$
$$= \frac{1}{2}[\alpha - \mu, \alpha + \mu],$$

(since $[\alpha, \mu] = [\mu, \alpha]$ for any 1-forms α, μ) so $dA = -\frac{1}{2}[A, B], B = \alpha + \mu$ which implies the condition on the $d\alpha_i$. The Frobenius theorem now says that every $(x, g) \in X \times G$ has a neighborhood U and a closed submanifold Z of U, containing (x, g), whose tangent space at every $(x', g') \in Z$ is $D_{x',g'}$.

Now re-topologize $X \times G$ so that it becomes a manifold of dimension of $\dim X$ with these Z as open sets. Noting that the projection pr_1 of $D_{x,g}$ to $T_x X$ is an isomorphism we see that pr_1 makes the retopologized $X \times G$ a covering space of X. (It helps to use the left action of G on $X \times G$). Let $\tilde{X}$ be the connected component of $(x_0, e) = \tilde{x}_0 \in X \times G$. Then $\tilde{X} \to X$ is a covering space and $pr_2|_{\tilde{X}} : \tilde{X} \to G$ is a map $Y : \tilde{X} \to G$ satisfying the required conditions. $\square$

Exercise 1   Show that in general there is no global map $Y : X \to G$ with $Y^*\mu = \alpha$ by finding a counterexample - a (very small) manifold X, a 1-form α on it, and a (very small) $G, \mathfrak{g}$, such that there is no $Y : X \to G$ as above.

Question The above proof seems not to have used fully the assumption that G is a Lie group. Can we then state a more general theorem?

Exercise 2   Prove directly that if V_1, V_2 are vector fields on $X \times G$ with $V_i(x, g) \in D_{x,g}$ for every (x, g) then $[V_1, V_2]$ satisfies the same condition.

1.3   Program

We will (following K.T.Chen) construct a special Lie algebra $\mathfrak{g}$ and a $\mathfrak{g}$-valued 1-form θ ("Chen's connection") for a manifold X - to simplify and make the construction canonical we assume X is compact Kahler.

As motivation, recall that for any simplicial complex X, the fundamental group can be defined as having generators corresponding to closed edge paths and relations corresponding to 2-dimensional simplices: thus only dimensions 1 and 2 are involved. Here, the Lie algebras $\mathfrak{g}$ will be defined as the free Lie algebra generated by the vector space $H_1(X;\mathbb{R}) = H_1$, modulo relations $\subset [H_1, H_1]$ obtained from the *reduced* diagonal map

$$\bar{\Delta}_* : H_2(X;\mathbb{R}) \to H_2(X \times X;\mathbb{R}) = (H_2 \otimes \mathbb{R}) \oplus (\mathbb{R} \otimes H_2) \oplus (H_1 \otimes H_1)$$
$$\to H_1 \otimes H_1.$$

The image $\bar{\Delta}_*(H_2) \subset H_1 \otimes H_1$ is in fact contained in the skew-symmetric elements $\sum a \otimes b - b \otimes a$ which we identify with $[H_1, H_1]$ in the free Lie algebra. There will be a number of technical points involved; $\mathfrak{g}$ will in general be an infinite dimensional Lie algebra and so θ will be some kind of infinite series. Similarly its "Lie group" G will be an inverse limit of finite dimensional LIe group.

The result will involve also some other associative algebras and another Lie algebra associated to $\pi_1 = \pi_1(X, x_0)$. We refer for the following to [Quillen] and [Lazard]. First, we form the associative algebra $\mathbb{R}\pi_1 = $ group algebra of π_1 with coefficients $\mathbb{R}$, then a descendng sequence of 2-sided ideals in $\mathbb{R}\pi_1$:

$$I = \text{augmentation ideal} \supset I^2 \supset I^3 \supset \cdots$$

and consider the associated graded algebra

$$Gr(\mathbb{R}\pi_1) = (\mathbb{R}\pi_1/I) \oplus (I/I^2) \oplus \cdots \oplus (I^n/I^{n+1}) \oplus \cdots$$

which is furthermore a Hopf algebra.

Next we consider the group π_1 itself and a sequence of normal subgroups (the lower central series); to indicate groups generated by commutators use round brackets; $(\pi_1, \pi_1), ((\pi_1, \pi_1), \pi_1)$,etc.

$$\pi_1 \supset (\pi_1, \pi_1) \supset ((\pi_1, \pi_1), \pi_1) \supset \cdots \supset \pi_1^{(n)} \supset \pi_1^{(n+1)} \supset \cdots$$

where $\pi_1^{(1)} = \pi_1$ and $\pi_1^{(n+1)} = (\pi_1^{(n)}, \pi_1)$. The quotients $\pi_1^{(n)}/\pi_1^{(n+1)}$ are abelian groups and the commutator operation in π_1 induces a map (a Lie bracket now)

$$[\,,\,] : \pi_1^{(n)}/\pi_1^{(n+1)} \otimes \pi_1^{(m)}/\pi_1^{(m+1)} \to \pi_1^{(n+m)}/\pi_1^{(n+m+1)},$$

which makes the associated graded abelian group $gr\pi_1 = \bigoplus_{n=1}^{\infty} \pi_1^{(n)}/\pi_1^{(n+1)}$ into a Lie algebra(we will tensor with $\mathbb{R}$).

Associated with the Lie algebras $\mathfrak{g}$ and $(gr\pi_1) \otimes \mathbb{R}$ are their enveloping associative algebras $U(\mathfrak{g})$ and $U(gr\pi_1 \otimes \mathbb{R})$ (also Hopf algebras). The group G associated to $\mathfrak{g}$ will be contained in a completion $U(\mathfrak{g})^{\wedge}$ of $U(\mathfrak{g})$. After all these algebras have been defined we will have:

1) The Chen connection θ, which gives a homomorphism $\pi_1 \to G$ of groups, will also induce a homomorphism

$$II : Gr(\mathbb{R}\pi_1) \to U(\mathfrak{g})$$

of Hopf algebras.

2) (As shown by Quillen) for any group π_1, there is a homomorphism (of Hopf algebras)

$$Q : U(Gr\pi_1 \otimes \mathbb{R}) \to Gr(\mathbb{R}\pi_1).$$

3) Both Q and II are isomorphisms

$$U(Gr(\pi_1) \otimes \mathbb{R}) \overset{Q}{\to} Gr(\mathbb{R}\pi_1) \overset{II}{\to} U(\mathfrak{g})$$

(Quillen showed Q is an isomorphism in general; the proof here will consider only compact Kahler X).

1.4   Lie algebras

We begin this program by reviewing the universal enveloping algebra $U(\mathfrak{g})$ of a Lie algebra $\mathfrak{g}$. Analytic definition of $U(\mathfrak{g})$ where G is a (finite dimensional) Lie group: we recall that the elements of $\mathfrak{g}$ are first order differential operators on G

$$\sum f_i(x_1, \cdots, x_n) \frac{\partial}{\partial x_i}$$

in local coordinates which are invariant under left translation. $\mathfrak{g}$ is closed under $[,]$ but not under usual (associative) product $\xi\eta$ (= second order differential operator). So we consider $U(\mathfrak{g})$ = the associative algebra of differential operators generated by $\mathfrak{g}$= all left invariant differential operators of all orders. However we will give a more algebraic construction of $U(\mathfrak{g})$

(valid over any field); $U(\mathfrak{g})$ will then be an associative algebra (with unit element 1) containing $\mathfrak{g}$ (as a sub-Lie algebra) and characterized by the following universal property: given any associative algebra A and any Lie algebra homomorphism $h : \mathfrak{g} \to A$, there is a unique extension of h to an associative algebra homomorphism of $U(\mathfrak{g})$ to A, agreeing with h on $\mathfrak{g}$ and taking 1 to 1.

To construct $U(\mathfrak{g})$ we first construct the tensor algebra

$$T(\mathfrak{g}) = \mathbb{R}1 \oplus \mathfrak{g} \oplus (\mathfrak{g} \otimes \mathfrak{g}) \oplus \cdots$$

on the vector space $\mathfrak{g}$, then factor out the 2-sided ideal $R \subset T(\mathfrak{g})$ generated by all elements of the following form: for $x, y \in \mathfrak{g}$, $r = x \otimes y - y \otimes x - [x, y]$ ($\in \mathfrak{g} \otimes \mathfrak{g} \oplus \mathfrak{g}$) is required to be in R. Thus $U(\mathfrak{g}) = T(\mathfrak{g})/R$ and it is easy to see that $\mathfrak{g} \to T(\mathfrak{g}) \to U(\mathfrak{g})$ is 1-1. The universal property of $U(\mathfrak{g})$ follows from a universal property of $T(\mathfrak{g})$.

The Poincarè-Birkhoff-Witt theorem describes $U(\mathfrak{g})$ as follows: if $v_1, v_2, \cdots$ is any vector space basis of $\mathfrak{g}$, then the monomials: $1, v_i, v_i v_j (i \leq j), \cdots, v_{i_1} v_{i_2} \cdots v_{i_k} (i_1 \leq \cdots \leq i_k)$ are a vector space basis for $U(\mathfrak{g})$. In this construction, $\mathfrak{g}$ need not be finite dimensional. In particular we may take $\mathfrak{g}$ to be the free Lie algebra generated by a vector space V:

$$\mathfrak{g} = V \oplus [V, V] \oplus [[V, V], V] \oplus \cdots$$

and find that $U(\mathfrak{g}) = T(V) =$ free associative algebra on V.

An important property of $U(\mathfrak{g})$ is that there is an associative algebra homomorphism

$$\Delta : U(\mathfrak{g}) \to U(\mathfrak{g} \oplus \mathfrak{g}) \xrightarrow{\text{(natural isom.)}} U(\mathfrak{g}) \otimes U(\mathfrak{g}). \qquad (1.5)$$

For $x \in \mathfrak{g}$, $x \mapsto x \oplus x \mapsto x \otimes 1 + 1 \otimes x$, $\Delta(1) = 1 \otimes 1$. For $x_1, x_2, \cdots x_n \in \mathfrak{g}$,

$$\begin{aligned}
\Delta(x_1 x_2 \cdots x_n) =& x_1 x_2 \cdots x_n \otimes 1 + 1 \otimes x_1 \cdots x_n \\
& + \sum x_{i_1} \cdots x_{i_k} \otimes x_{j_1} \cdots x_{j_l} \quad (k + l = n).
\end{aligned} \qquad (1.6)$$

Thinking of $\mathfrak{g}$ as vector fields ξ on a Lie group G, Δ is the Lie algebra homomorphism induced by the diagonal homomorphism

$$G \to G \times G, \quad g \mapsto (g, g).$$

The associative algebra $U(\mathfrak{g})$ together with $\Delta : U(\mathfrak{g}) \to U(\mathfrak{g}) \otimes U(\mathfrak{g})$ is an example of a *Hopf algebra* (defined later). The elements of $\mathfrak{g}$ are *primitive* in $U(\mathfrak{g})$: $\Delta(x) = x \otimes 1 + 1 \otimes x$, and in fact

$$\mathfrak{g} = P(U(\mathfrak{g})) = \{u \in U(\mathfrak{g}) : \Delta(u) = u \otimes 1 + 1 \otimes u\}.$$
$$(= \text{all primitive elements}).$$

Exercise 3 Let $I_\mathfrak{g}$ be an ideal in the Lie algebra $\mathfrak{g}$, *i.e.*, $x \in I_\mathfrak{g}, y \in \mathfrak{g} \Rightarrow$ $[x, y] \in I_\mathfrak{g}$. Thus $\mathfrak{g}/I_\mathfrak{g}$ is again a Lie algebra. Considering $I_\mathfrak{g} \subset \mathfrak{g} \subset U(\mathfrak{g})$, let R be the 2-sided associative algebra ideal in $U(\mathfrak{g})$ generated by $I_\mathfrak{g}$ (so, $R =$ all $\sum u_i x_i v_i, u_i, v_i \in U(\mathfrak{g}), x_i \in I_\mathfrak{g}$). Then $U(\mathfrak{g}/I_\mathfrak{g})$ is naturally isomorphic to $U(\mathfrak{g})/R$. (Use the universal property).

In the Chen description of $\pi_1(X, x_0)$ we will use the following Lie algebra $\mathfrak{g}$: consider the free Lie algebra L generated by $H_1(X; \mathbb{R})$ namely, $L = H_1 \oplus [H_1, H_1] \oplus \cdots$. Note that $[H_1, H_1] = \{\sum a \otimes b - b \otimes a : a, b \in H_1\}$. Furthermore consider the diagonal map $\Delta : X \to X \times X$ and the following sequence of $\mathbb{R}$-linear maps

$$\bar{\Delta}_* : H_2(X) \xrightarrow{\Delta_*} H_2(X \times X) \xrightarrow{\sim} (H_1 \otimes H_1) \oplus (H_2 \otimes H_0 \oplus H_0 \otimes H_2)$$
$$\xrightarrow{project} H_1 \otimes H_1.$$

The image of $\bar{\Delta}_*$ actually lies in $[H_1, H_1]$ =the skew-symmetric elements of $H_1 \otimes H_1$ (dually, the cup product map $H^1 \otimes H^1 \to H^2$ is skew-symmetric).

Let now I be the Lie algebra ideal in L(= free Lie algebra on H_1) generated by $\bar{\Delta}_*(H_2(X; \mathbb{R}))$. We define $\mathfrak{g} = L/I$=Lie algebra over $\mathbb{R}$.

The Hopf algebras we will use are associative algebras H over $\mathbb{R}$ with unit element 1 and algebra homomorphism $\epsilon : H \to \mathbb{R}$ $(\epsilon(1) = 1)$ so that ker ϵ, denoted $\bar{H}$ is a 2-sided ideal and $H = \mathbb{R}1 \oplus \bar{H}$ as $\mathbb{R}$-vector space. The Hopf algebra structure is an associative algebra homomorphism

$$\Delta : H \to H \otimes H, \quad \Delta(1) = 1 \otimes 1,$$

such that if $h \in \bar{H}$ then

$$\Delta(h) = h \otimes 1 + 1 \otimes h + \sum h_i' \otimes h_i'' \quad (h_i', h_i'' \in \bar{H}).$$

Thus

$$\Delta(\bar{H}) \subset (\bar{H} \otimes \mathbb{R}1) \oplus (\mathbb{R}1 \otimes \bar{H}) \oplus (\bar{H} \otimes \bar{H}) = \overline{H \otimes H}.$$

The main examples of Hopf algebras we use are:

1. $H = \mathbb{R}\pi$, π any group. (In fact, π could be just a semigroup with identity e, *i.e.*, a "monoid"). Δ is just the homomorphism

$$\Delta : \mathbb{R}\pi \to \mathbb{R}(\pi \times \pi) = \mathbb{R}\pi \otimes \mathbb{R}\pi$$

given by the diagonal homomorphism $\Delta : \pi \to \pi \times \pi$,

$$\Delta(g) = (g,g) \leftrightarrow g \otimes g \text{ in } \mathbb{R}\pi \otimes \mathbb{R}\pi$$
$$\Delta(\textstyle\sum a_g g) = \sum a_g g \otimes g \quad (a_g \in \mathbb{R}, \, g \in \pi).$$

2. Similarly, $U(\mathfrak{g})$, for a Lie algebra $\mathfrak{g}$ has diagonal map $\Delta : U(\mathfrak{g}) \to U(\mathfrak{g}) \otimes U(\mathfrak{g})$ given by the diagonal homomorphism

$$\mathfrak{g} \to \mathfrak{g} \oplus \mathfrak{g}, \quad x \mapsto (x,x).$$

So, $\Delta(x) = x \otimes 1 + 1 \otimes x$ for $x \in \mathfrak{g}$.

3. If H is a Hopf algebra with map Δ, a 2-sided associative algebra ideal $I \subset \bar{H}$ is called a Hopf ideal if

$$\Delta(I) \subset I \otimes H + H \otimes I \subset H \otimes H.$$

Then H/I is again a Hopf algebra (with diagonal map induced by Δ). Note that for all $n \geq 1$,

$$\Delta(I^n) \subset H \otimes I^n + I \otimes I^{n-1} + \cdots + I^n \otimes H = \Delta(I)^n.$$

Thus the associated graded associative algebra

$$\oplus_{n \geq 0} I^n / I^{n+1} = H/I \oplus I/I^2 \oplus \cdots$$

is also a Hopf algebra if I is a Hopf ideal.

Definition 1.1   1. An element $x \in \bar{H}$ is called *primitive* if

$$\Delta(x) = x \otimes 1 + 1 \otimes x.$$

The primitive elements are a Lie algebra under $[,]$, denoted $P(H)$.

2. An element $g \in H$ will be called "group-like" if it is invertible and satisfies $\Delta(g) = g \otimes g$. The group-like elements are a group under multiplication.

Note that $g = 1 + \bar{g}, \bar{g} \in \bar{H}$, if g is group-like.

Example 1.1   *a*) The 2-sided ideal I in H generated by a collection of primitive elements x is a Hopf ideal.

b) The primitive elements in $U(\mathfrak{g})$ are exactly $\mathfrak{g}$ (use the Poincarè-Birkhoff-Witt theorem to prove this!)

c) The group-like elements in $\mathbb{R}\pi$ are just π (for π a group). (Prove this!)

We look now at the process of *completing* certain Lie algebras and their enveloping algebras, to obtain something resembling formal power series, in which we can construct exponential and log functions.(See [Lazard]).

First, let L be the free Lie algebra on a vector space V (in our applications V will be finite dimensional). Then $U(L)$ is the tensor algebra (or free associative algebra) $T(V)$ and L is the Lie subalgebra of $T(V)$ generated by V;

$$L = V \oplus [V, V] \oplus [[V, V], V] \oplus \cdots = \oplus_{i=1}^{\infty} L_i$$

where the subspaces $L_1 = V$, $L_{i+1} = [L_i, V]$ satisfy $[L_i, L_j] \subset L_{i+j}$. L is called a graded Lie algebra.

For each n, $\oplus_{i \geq n} L_i$ denoted $L_{\geq n}$ is a Lie ideal. Thus we have a sequence of onto homomorphism of Lie algebras

$$\rightarrow L/L_{\geq n} \rightarrow L/L_{\geq n-1} \rightarrow \cdots \rightarrow L/L_{\geq 2} = V \rightarrow 0$$

and the inverse limit is a Lie algebra $L^{\wedge}$ whose elements may be written as infinite series

$$l^{\wedge} = l_1 + l_2 + l_3 + \cdots, l_i \in L_i.$$

Next consider a graded ideal I_L in L;

$$I_L = (I_L \cap L_1) \oplus (I_L \cap L_2) \oplus \cdots.$$

Then L/I_L is again a graded Lie algebra $\mathfrak{g}$ and we can form the completion $\mathfrak{g}^{\wedge}$;

$$\mathfrak{g} = \mathfrak{g}_1 \oplus \mathfrak{g}_2 \oplus \cdots, \quad \mathfrak{g}_i = L_i/I_L \cap L_i$$

and

$$\mathfrak{g}^{\wedge} = \varprojlim(\mathfrak{g}/\mathfrak{g}_{\geq n})$$

with elements $x^{\wedge} = x_1 + x_2 + \cdots$ (again formal infinite series).

Since $\mathfrak{g} = L/I_L$, $U(\mathfrak{g}) = U(L)/R$, $R=$ 2-sided ideal in $U(L)$ generated by I_L. So R is again a graded, or homogeneous ideal and $U(\mathfrak{g}) = U(L)/R$ is a graded associative algebra generated as associative algebra by its degree 1 elements $\mathfrak{g}_1 = V/(I_L \cap V)$. For simplicity we will assume $I_L \cap V = (0)$ so $\mathfrak{g}_1 = V = L_1$ (as will be the case in our examples).

Now consider in $U(\mathfrak{g})$ the ideal of all elements of degree $\geq n$, which is just $(V_1)^n = (\mathfrak{g})^n = U(\mathfrak{g})_{\geq n}$ (note this is larger than the ideal generated by

$\mathfrak{g}_{\geq n}$). Then the inverse limit of the $U(\mathfrak{g})/U(\mathfrak{g})_{\geq n}$ will be denoted $U(\mathfrak{g})^{\wedge}$: it is an associative algebra with diagonal map homomorphism

$$\Delta^{\wedge} : U(\mathfrak{g})^{\wedge} \to U(\mathfrak{g} \oplus \mathfrak{g})^{\wedge} = (U(\mathfrak{g}) \otimes U(\mathfrak{g}))^{\wedge}.$$

Since $(V_1)^n \cap \mathfrak{g} = \mathfrak{g}_{\geq n}$, $\mathfrak{g}^{\wedge}$ embeds in $U(\mathfrak{g})^{\wedge}$ as Lie subalgebra: both of these have elements which are all the infinite series, with terms in $\mathfrak{g}_i$ or V_1^i respectively.

Theorem 1.2   *1. The primitive elements of $U(\mathfrak{g})^{\wedge}$ with respect to $\Delta^{\wedge}$ are just the elements of $\mathfrak{g}^{\wedge}$.*

2. Let $x \in U(\mathfrak{g})^{\wedge}$ have constant term 0, and define $\exp x = 1 + x + x^2/2! + \cdots$, an element of $U(\mathfrak{g})^{\wedge}$. Similarly define $\log(1+x) = x - x^2/2 + x^3/3 - \cdots$. Then: a) if $p \in U(\mathfrak{g})^{\wedge}$ is primitive, then $\exp(p) = g \in U(\mathfrak{g})^{\wedge}$ is group-like:

$$\Delta^{\wedge}(\exp p) = \exp p \otimes \exp p \in (U(\mathfrak{g}) \otimes U(\mathfrak{g}))^{\wedge}$$

(and $\exp p$ has constant term 1 and is invertible).

b) If $g = 1 + x \in U(\mathfrak{g})^{\wedge}$ is group-like then $\log g$ is primitive (and so is in $\mathfrak{g}^{\wedge}$).

3. We have bijections which are inverse to each other

$$Primitives = \mathfrak{g}^{\wedge} \; \underset{\log}{\overset{\exp}{\longleftarrow\!\!\!\longrightarrow}} \; G = group\text{-}like \; elements \; of \; U(\mathfrak{g})^{\wedge}.$$

Proof: 1. If $p = p_1 + p_2 + \cdots$, $(p_i \in U(\mathfrak{g})_i$ then

$$\Delta^{\wedge}(p) = p \otimes 1 + 1 \otimes p$$

if and only if

$$\Delta(p_i) = p_i \otimes 1 + 1 \otimes p_i \in U(\mathfrak{g}) \otimes U(\mathfrak{g})$$

for each i, so $p_i \in \mathfrak{g}_i$ (primitives of $U(\mathfrak{g}) = \mathfrak{g}$).

2. *a)* $\Delta^{\wedge}(p) = p \otimes 1 + 1 \otimes p$ implies

$$\begin{aligned} \Delta^{\wedge}(\exp p) &= \exp(\Delta^{\wedge}(p)) = \exp(p \otimes 1 + 1 \otimes p) \\ &= \exp(p \otimes 1)\exp(1 \otimes p) \quad \text{as } p \otimes 1, 1 \otimes p \text{ commute} \\ &= \exp p \otimes \exp p, \end{aligned}$$

so $\exp p = g$ is group-like.

b) If $\Delta^\wedge(g) = g \otimes g$ $(g = 1 + x)$ then

$$\Delta^\wedge(\log g) = \log(\Delta^\wedge(g)) = \log((g \otimes 1)(1 \otimes g))$$
$$= (\log g) \otimes 1 + 1 \otimes (\log g),$$

so $\log g$ is primitive. $\square$

In the discussion above we can replace $\mathfrak{g}$ by $\mathfrak{g}/\mathfrak{g}_{\geq n}$ which is an $(n-1)$ step nilpotent and graded Lie algebra. Then in the completion $U(\mathfrak{g}/\mathfrak{g}_{\geq n})^\wedge$ we have $(\mathfrak{g}/\mathfrak{g}_{\geq n})^\wedge = \mathfrak{g}/\mathfrak{g}_{\geq n}$ (each series actually has only a finite number $< n$ of terms) and the corresponding group $\exp(\mathfrak{g}/\mathfrak{g}_{\geq n})$ has multiplication defined purely by the bracket in $\mathfrak{g}/\mathfrak{g}_{\geq n}$:

$$\exp x \exp y = \exp(x + y + \frac{1}{2}[x,y] + \cdots).$$

Suppose further that $\mathfrak{g}$ is finite dimensional over $\mathbb{R}$: then $\mathfrak{g}/\mathfrak{g}_{\geq n}$ is also finite dimensional and the corresponding group is diffeomorphic to $\mathfrak{g}/\mathfrak{g}_{\geq n}$ via $\exp, \log$ and is a nilpotent Lie group.

1.5   Chen's Lie algebra and connection

Now let us consider a connected manifold X with finite dimensional $H_1(X; \mathbb{R}) = H_1$ and define $\mathfrak{g}$ as (free Lie algebra on H_1)/ ideal $(\bar{\Delta}_* H_2)$. Let us consider any $\mathfrak{g}^\wedge$-valued 1-form θ on X, by which we mean an infinite series

$$\theta = \theta_1 + \theta_2 + \cdots$$

with $\theta_i = $1-form on X with values in $\mathfrak{g}$. We will assume that

$$d\theta + \frac{1}{2}[\theta, \theta] = 0,$$

and that θ_1 has the following special form: choose any basis $h_{1,i}$ of H_1 over $\mathbb{R}$, and let α_i^1 be closed 1-forms on X whose cohomology classes are dual to the $h_{1,i}$, i.e.,

$$\int_{h_i} \alpha_j^1 = \delta_{i,j}.$$

Then we assume $\theta_1 = \sum_i \alpha_i^1 \otimes h_{1,i}$.

If we fix $n > 1$ and consider $\mathfrak{g}/\mathfrak{g}_{\geq n} = \mathfrak{g}_{(n)}$, we get a "homomorphic image"

$$\theta_{(n)} = \theta \mod \mathfrak{g}_{\geq n}$$

which again is a flat connection and gives a homomorphism which we denote

$$II : \pi_1(X, x_0) \to G_{(n)} = \text{Lie group of } \mathfrak{g}/\mathfrak{g}_{\geq n}$$

explicitly given by the iterated integral series whose degree 1 term, for an element $\gamma \in \pi_1(X, x_0)$ is just $\sum_i (\int_\gamma \alpha_i^1) h_{1,i} = $ image of γ in $H_1(X; \mathbb{R}) = \pi_1/[\pi_1, \pi_1] \otimes \mathbb{R}$.

The group homomorphism given by $\theta_{(n)}$:

$$h_{(n)} : \pi_1(X, x_0) \to G_{(n)} \subset U(\mathfrak{g}/\mathfrak{g}_{\geq n})^\wedge$$

(also denoted II) extends to an associative algebra homomorphism, again denoted generically as II

$$\mathbb{R}\pi_1 \to U(\mathfrak{g}_{(n)})^\wedge$$

which is compatible with the diagonal maps Δ, $\Delta^\wedge$ since group elements in π_1 go into group-like elements in $G(n)$: it is a "Hopf homomorphism" $h_{(n)}$, which is compatible also with homomorphisms given by decreasing n. Thus, $h_{(n)}$ commutes with homomorphisms to $\mathbb{R}$ defining the augmentation ideals, and so induces also a homomorphism of the completion $(\mathbb{R}\pi_1)^\wedge \to U(\mathfrak{g}_{(n)})^\wedge$ and a homomorphism of the associated graded algebras:

$$Grh_{(n)} : Gr(\mathbb{R}\pi_1) \to Gr(U(\mathfrak{g}_{(n)})^\wedge) = U(\mathfrak{g}_{(n)})$$

(since $U(\mathfrak{g}(n))$ is a graded algebra).

This last homomorphism induces just the natural isomorphism on the degree 1 elements

$$(I\pi_1)/(I\pi_1)^2 = \pi_1/(\pi_1, \pi_1) \otimes \mathbb{R} \to H_1(X, \mathbb{R}) = \mathfrak{g}_1$$

which are generators, and so $Grh_{(n)}$ is surjective. The $Grh_{(n)}$ for increasing n are compatible and define a Hopf homomorphism(surjective)

$$II = Grh : Gr\mathbb{R}\pi_1 \to U(\mathfrak{g}).$$

We also have surjective homomorphisms $(\mathbb{R}\pi_1)^\wedge \to U(\mathfrak{g})^\wedge$ and $\mathbb{R}\pi_1/(I\pi_1)^n \to U(\mathfrak{g})/(\mathfrak{g})^n$ for all n. We will prove all of these are isomorphisms.

Before proving this, we want to construct connections θ satisfying the two conditions above. We will consider only compact oriented (connected) manifolds X satisfying a special condition on 1-forms and 2-forms, which will be satisfied by all compact complex Kahler manifolds (and in particular by all non-singular complex algebraic varieties in projective space). The assumption is that we are given $\mathbb{R}$-linear subspaces:

$$\mathcal{H}^1, C^1 \text{ of } A^1(X),$$

$$\mathcal{H}^2, E^2 \text{ of } A^2(X)$$

$(A^i(X) = \text{all real valued differentiable } i\text{-forms})$ satisfying

1) $\mathcal{H}^i \subset \ker d$ and $\mathcal{H}^i \to \ker d / dA^{i-1} = H^i_{DR}(X)$ is an isomorphism.

2) $d : A^1 \to A^2$ induces an isomorphism $C^1 \to E^2$

3a) $\mathcal{H}^1 \wedge \mathcal{H}^1 \subset \mathcal{H}^2 + E^2$

(3a just says that if $\alpha, \beta \in \mathcal{H}^1$ so that $\alpha \wedge \beta$ represents a cohomology class in H^2, then $\alpha \wedge \beta = \gamma + \eta$, where $\gamma \in \mathcal{H}^2$ represents this cohomology class and $\eta \in E^2 \subset dA^1$ so η is exact).

3b)

$$[(\mathcal{H}^1 + C^1) \wedge C^1] \cap \ker d \subset E^2.$$

We remark that 1, 2, 3a hold on any Riemannian (compact) manifold if we write $\mathcal{H}^p =$ harmonic p-forms

$$A^p = \mathcal{H}^p \oplus dA^{p-1} \oplus d^* A^{p+1}$$

and take

$$C^1 = d^* A^2 \ (= \text{"coexact" 1-forms}) , \quad E^2 = dA^1 = dC^1 = \text{exact 2-forms}.$$

However 3b requires that the metric be Kahler as well (we define this later). A basic theorem of K.T.Chen is:

Theorem 1.3   *(K.T.Chen) Given the subspaces*

$$\mathcal{H}^1, C^1 \subset A^1(X); \ \mathcal{H}^2, E^2 \subset A^2(X)$$

satisfying 1, 2, 3 above there is a unique

$$\theta = \theta_1 + \theta_2 + \cdots$$

(formal infinite series with $\theta_i \in A^1(X) \otimes \mathfrak{g}_i$), where $\mathfrak{g}$ is the free Lie algebra on H_1, modulo relations $\bar{\Delta}_*(H_2)$, satisfying the following 3 conditions

I. $\theta_1 \in \mathcal{H}^1 \otimes H_1$ and represents the identity map $H_1 \to H_1$ if we identify $\mathcal{H}^1$ with $H^1 = $ vector space dual of H_1. (We are assuming these spaces are finite-dimensional): $\theta_1 = \sum \alpha_i \otimes x_i$.

II. For $i \geq 2$, $\theta_i \in C^1 \otimes \mathfrak{g}_i$.

III. $d\theta + \frac{1}{2}[\theta, \theta] = 0$.

Proof: Existence is shown by constructing $\theta_1, \theta_2, \cdots$ inductively, starting with θ_1; the proof will also give uniqueness. So we start with any basis $x_1, x_2, \cdots$ of $H_1 = \mathfrak{g}_1$ and "dual" basis $\alpha_1, \alpha_2, \cdots$ of $\mathcal{H}^1$, i.e., $\int_{x_j} \alpha_i = \delta_{i,j}$, and take $\theta_1 = \sum_i \alpha_i \otimes x_i$.

Thus, $d\theta_1 = 0$, while III requires $d\theta_2 = -\frac{1}{2}[\theta_1, \theta_1]$. Here

$$[\theta_1, \theta_1] = \sum_{i,j} \alpha_i \wedge \alpha_j \otimes [x_i, x_j].$$

By 3a),

$$\alpha_i \wedge \alpha_j = \mathcal{H}(\alpha_i \wedge \alpha_j) + a_{ij}$$

where $\mathcal{H}(\alpha_i \wedge \alpha_j) \in \mathcal{H}^2$ and $a_{ij} \in E^2$; these last two elements are uniquely determined by $\alpha_i \wedge \alpha_j$ and $a_{ij} = -a_{ji}$ since $\alpha_i \wedge \alpha_j = -\alpha_j \wedge \alpha_i$.

By 2), $d: C^1 \to E^2$ being an isomorphism, there exists a unique $\alpha_{ij} \in C^1$ with

$$d\alpha_{ij} = -a_{ij} = \mathcal{H}(\alpha_i \wedge \alpha_j) - \alpha_i \wedge \alpha_j.$$

So $\alpha_{ji} = -\alpha_{ij}$.

Now the requirement III says that

$$
\begin{aligned}
d\theta_2 &= -\frac{1}{2}[\theta_1, \theta_1] \\
&= -\frac{1}{2} \sum_{i,j} \alpha_i \wedge \alpha_j \otimes [x_i, x_j] \\
&= -\frac{1}{2} \sum (\mathcal{H}(\alpha_i \wedge \alpha_j) - d\alpha_{ij}) \otimes [x_i, x_j].
\end{aligned}
$$

We will show that $\sum \mathcal{H}(\alpha_i \wedge \alpha_j) \otimes [x_i, x_j] = 0$ in $A^2 \otimes \mathfrak{g}_2$ so that the above equation reduces to

$$d\theta_2 = \frac{1}{2}\sum d\alpha_{ij} \otimes [x_i, x_j] \quad (d\alpha_{ij} \in E^2 = dA^1).$$

Since $d : C^1 \to E^2$ is an isomorphism, this equation for θ_2 has a unique solution in $C^1 \otimes \mathfrak{g}_2$, namely

$$\theta_2 = \frac{1}{2}\sum_{i,j} \alpha_{ij} \otimes [x_i, x_j].$$

Now we show that for the basis α_i of $\mathcal{H}^1$ dual to basis x_i of H_1 we have

$$\sum_{i,j} \mathcal{H}(\alpha_i \wedge \alpha_j) \otimes [x_i, x_j] = 0 \in \mathcal{H}^2 \otimes \mathfrak{g}_2$$

($\mathcal{H}(\alpha_i \wedge \alpha_j)$ = harmonic part of $\alpha_i \wedge \alpha_j$): the left hand side can be regarded as a linear transformation $H_2 \to \mathfrak{g}_2$ (identifying $\mathcal{H}^2$ and H^2) which takes a homology class $\tau_2 \in H_2$ to $\sum_{i,j} (z_2, \alpha_i \wedge \alpha_j)[x_i, x_j]$. (($z_2, \alpha_i \wedge \alpha_j$) means $\int_{z_2} \alpha_i \wedge \alpha_j = \int_{z_2} \mathcal{H}(\alpha_i \wedge \alpha_j)$.)

It is more convenient now to write

$$\sum \alpha_i \wedge \alpha_j \otimes [x_i, x_j] = 2 \sum_{i,j} \alpha_i \wedge \alpha_j \otimes x_i x_j$$

and show

$$\sum_{i,j} (z_2, \alpha_i \wedge \alpha_j) x_i x_j = 0.$$

But $\alpha_i \wedge \alpha_j = \Delta^*(\alpha_i \otimes \alpha_j)$, ($\alpha_i \otimes \alpha_j$ being a form on $X \times X$) so

$$(z_2, \alpha_i \wedge \alpha_j) = (z_2, \Delta^*(\alpha_i \otimes \alpha_j)) = (\Delta_*(z_2), \alpha_i \otimes \alpha_j)$$

and we have to show that $\sum_{i,j}(\alpha_i \otimes \alpha_j) \otimes x_i x_j$ vanishes on all $\Delta_*(z_2)$. But if we consider $x_i x_j$ as being the image of $x_i \otimes x_j$, which is in the free associative algebra on H_1, then $\sum(\alpha_i \otimes \alpha_j) \otimes (x_i \otimes x_j)$ is the linear transformations of $H_1 \otimes H_1$ to itself taking $h \otimes h'$ to $\sum \alpha_i(h)x_i \otimes \alpha_j(h')x_j = h \otimes h'$, *i.e.*, the identity linear transformation. Thus $\sum(\alpha_i \otimes \alpha_j) \otimes (x_i \otimes x_j)$ takes $\bar{\Delta}_*(z_2)$ to itself and $\sum \alpha_i \otimes \alpha_j \otimes x_i x_j$ takes $\bar{\Delta}_*(z_2)$ to its image in $\mathfrak{g}_2$, which was expressly designed to be 0.

Now for $\theta_n, n \geq 3$, we already have $\theta_1 \in \mathcal{H}^1 \otimes \mathfrak{g}_1$ and we assume we have also defined $\theta_i \in C^1 \otimes \mathfrak{g}_i$ for $1 < i < n$ satisfying: for $1 \leq j \leq n - 1$,

$d\theta_j + \sum_{i=1}^{j-1} \theta_i \wedge \theta_{j-1} = 0$. Consider

$$\sum_{i=1}^{n-1} \theta_i \wedge \theta_{n-i} \in (\mathcal{H}^1 + C^1) \wedge C^1 \otimes \mathfrak{g}_n.$$

We want to show that $d(\sum_{i=1}^{n-1} \theta_i \wedge \theta_{n-i}) = 0$.

Then assumption 3*b* will give us that the 2-form coefficients will be in $((\mathcal{H}^1 + C^1) \wedge C^1) \cap \ker d = E^2$, and E^2 is isomorphic to C^1 via d, so there will exist a unique $\theta_n \in C^1 \otimes \mathfrak{g}_n$ such that $d\theta_n = -\sum_{i=1}^{n-1} \theta_i \wedge \theta_{n-i}$ which satisfies condition III for θ_n. But,

$$d(\sum_{i=1}^{n-1} \theta_i \wedge \theta_{n-i}) = \sum_{1}^{n-1} d\theta_i \wedge \theta_{n-i} - \theta_i \wedge d\theta_{n-i}$$

$$= \text{a sum of terms } \pm \theta_a \wedge \theta_b \wedge \theta_c.$$

Consider triples (p, q, r) with $p, q, r \geq 1$ and $p + q + r = n$. Then $\theta_p \wedge \theta_q \wedge \theta_r$ occurs in this sum: once with coefficient -1, in $(d\theta_{p+q}) \wedge \theta_r$ once with coefficient +1, in $-\theta_p \wedge d\theta_{q+r}$ and nowhere else. So the sum is 0 concluding the proof of the theorem.□

We come back to our program of constructing homomorphisms of Lie algebras, associative algebras, and Hopf algebras which we will show are isomorphisms.

First of all, from the connection θ we have homomorphisms, denoted II(=iterated integral) of groups:

$$\pi_1 \to G = \text{group-like elements in } U(\mathfrak{g})^\wedge$$

of Hopf algebras:

$$\mathbb{R}\pi_1 \to (\mathbb{R}\pi_1)^\wedge \to U(\mathfrak{g})^\wedge$$

of graded algebras:

$$Gr\mathbb{R}\pi_1 \to U(\mathfrak{g}).$$

1.6   Some work of Quillen

Next, following Quillen and Lazard, for any group π we consider its descending central series and associated graded abelian group $Gr\pi$, which is a Lie algebra (using group commutator in π).

The function $q : \pi \to \overline{\mathbb{R}\pi}$, $q(\gamma) = \gamma - 1$ induces a homomorphism Q of a Lie algebra to an associative algebra

$$Gr\pi \to Gr\mathbb{R}\pi$$

and so a homomorphism

$$Q : U(Gr\pi \otimes \mathbb{R}) \to Gr(\mathbb{R}\pi)$$

of Hopf algebras.

We will now give the details of Q. We will use the following construction studied in [Quillen]: for any group π, form the descending central series $((\,,\,)$ denoting group commutator)

$$\pi^{(1)} = \pi \supset \pi^{(2)} = (\pi, \pi) \supset \cdots \supset \pi^{(n+1)} = (\pi^{(n)}, \pi) \supset \cdots$$

and consider the graded abelian group $Gr(\pi) = \oplus_{n \geq 1} \pi^{(n)}/\pi^{(n+1)}$ as a Lie algebra using group commutator in π (see [Lazard]).

We want to construct a Lie algebra homomorphism

$$(Gr\pi) \otimes \mathbb{R} \to Gr(\overline{\mathbb{R}\pi}) = \oplus_{n \geq 1} (\overline{\mathbb{R}\pi})^n/(\overline{\mathbb{R}\pi})^{n+1},$$

where $[,]$ in $Gr(\overline{\mathbb{R}\pi})$ is given by commutator in this associative algebra. We start with the function

$$q : \pi \to \overline{\mathbb{R}\pi}, \quad q(\gamma) = \gamma - 1$$

inducing the abelian group homomorphism

$$\pi^{(1)}/\pi^{(2)} \to \overline{\mathbb{R}\pi}/(\overline{\mathbb{R}\pi})^2$$
$$(\gamma\pi^{(2)}) \to q(\gamma) + (\overline{\mathbb{R}\pi})^2.$$

This induces an isomorphism over $\mathbb{R}$:

$$\pi^{(1)}/\pi^{(2)} \otimes \mathbb{R} \to \overline{\mathbb{R}\pi}/(\overline{\mathbb{R}\pi})^2.$$

These identities will be used:
a) $q(\gamma_1\gamma_2) = \gamma_1\gamma_2 - 1 = (\gamma_1 - 1) + (\gamma_2 - 1) + (\gamma_1 - 1)(\gamma_2 - 1)$,
b)

$$q(\gamma_1\gamma_2\gamma_1^{-1}\gamma_2^{-1}) = ((\gamma_1 - 1)(\gamma_2 - 1) - (\gamma_2 - 1)(\gamma_1 - 1))\gamma_1^{-1}\gamma_2^{-1}$$
$$= [(\gamma_1 - 1), (\gamma_2 - 1)](1 + (\gamma_1^{-1}\gamma_2^{-1} - 1))$$
$$= \text{elements of } I^2$$
$$= [(\gamma_1 - 1), (\gamma_2 - 1)] \mod I^3.$$

Next, assume inductively that for a given $n \geq 2$, $q(\pi_1^{(n)}) \in \overline{\mathbb{R}\pi}^n$ and if $\gamma_1 \in \pi^{(n-1)}$, $\gamma_2 \in \pi$ then

$$q((\gamma_1, \gamma_2)) = [q(\gamma_1), q(\gamma_2)] \mod (\overline{\mathbb{R}\pi})^{(n+1)}.$$

Using $a)$ and then $b)$ we prove that

$$q : \pi_1^{(n)} \to \overline{\mathbb{R}\pi}^n / \overline{\mathbb{R}\pi}^{n+1}$$

is a group homomorphism, induces

$$q : \pi_1^{(n)} / \pi_1^{(n+1)} \to \overline{\mathbb{R}\pi}^n / \overline{\mathbb{R}\pi}^{n+1}$$

also a homomorphism, and if $\gamma_1 \in \pi_1^{(n)}$, $\gamma_2 \in \pi$ then $q((\gamma_1, \gamma_2))$ belongs to $\overline{\mathbb{R}\pi}^{n+1}$ and $= [\gamma_1 - 1, \gamma_2 - 1] \mod \overline{\mathbb{R}\pi}^{n+2}$. Thus $q(\pi_1^{(n+1)}) \subset \overline{\mathbb{R}\pi}^{n+1}$, q induces $\pi_1^{(n)} / \pi_1^{(n+1)} \to \overline{\mathbb{R}\pi}^n / \overline{\mathbb{R}\pi}^{n+1}$ and takes group commutator $(,) \in (\pi_1^{(n)}, \pi_1)$ into algebra commutator $[,] \in \overline{\mathbb{R}\pi}^{n+1} / \overline{\mathbb{R}\pi}^{n+2}$. Finally, one can show that q induces a graded Lie algebra homomorphism $Gr(\pi) \to Gr(\mathbb{R}\pi)$ which is an isomorphism on $\pi/\pi^{(2)} \otimes \mathbb{R} \to \overline{\mathbb{R}\pi}/\overline{\mathbb{R}\pi}^2$.

Thus one gets a homomorphism of associative algebras and even of Hopf algebras

$$Q : U((Gr\pi) \otimes \mathbb{R}) \to Gr(\mathbb{R}\pi).$$

Also (see [Quillen]), $Gr(\mathbb{R}\pi)$ is the universal enveloping algebra $U(P(Gr\mathbb{R}\pi))$, where P denotes the Lie algebra of all primitive elements (the Hopf structure in $Gr(\mathbb{R}\pi)$ arises from that in $\mathbb{R}\pi$).

The Hopf algebra map Q thus induces a Lie algebra map

$$Q : (Gr\pi) \otimes \mathbb{R} \to P(Gr\mathbb{R}\pi).$$

Similarly II induces Lie algebra map

$$II : P(Gr(\mathbb{R}\pi)) \to \mathfrak{g} = Gr(\mathfrak{g}^\wedge) = P(U(\mathfrak{g})).$$

We aim to show that all these Q, II are isomorphisms.

1.7 Group homology

We will use some homology theory of discrete groups $\pi = \pi_1(X)$. Let $B\pi$ be the classifying space of π; it is a $K(\pi, 1)$ space, *i.e.*, its fundamental group is π and its higher homotopy groups π_n for $n \geq 2$ vanish. The homology groups with coefficients in any $\mathbb{R}\pi$ module M, $H_i(B\pi, M)$ are denoted $H_i(\pi, M)$ (and depend only on π). For $i = 0$, $H_0(\pi, M) = M/\mathbb{R}\pi M$.

For the (connected) manifold X and $\pi = \pi_1(X, x_0)$, we construct $Y = B\pi_1$ as $Y = X \cup$ cells of dimension ≥ 3. Thus the inclusion $X \to Y$ induces an isomorphism $\pi_1(X) \to \pi_1(Y)$. Up to homotopy we can replace the inclusion $X \subset Y$ by a fibration $p : \tilde{X} \to Y$, with $X \to \tilde{X}$ a homotopy equivalence, with fiber $p^{-1}(y_0) = F$ (where $y_0 =$ base point of Y). Notationally we will write X instead of $\tilde{X}$ (we know $H_i(X) = H_i(\tilde{X})$ and $\pi_i(X) = \pi_i(\tilde{X})$.)

For any fibration $F \to X \to Y$ as above (a "Serre fibration"), we have some exact sequences of homology groups, called exact sequences of low order terms in a spectral sequence: see H. Cartan and S. Eilenberg's book "Homological Algebra" (p. 328 Th.5.12a with $n = 1$). In these $H_1(F) = H_1(F; \mathbb{R})$ is a $\pi_1 = \pi_1(X, x_0)$ module and $H_1(F)_{\pi_1}$ means $H_1(F)/\overline{\mathbb{R}\pi_1}H_1(F)$.

The exact sequences are

$$H_2(X) \to H_2(Y) \to H_1(F)_{\pi_1} \to H_1(X) \to H_1(Y) \to 0. \qquad (1.7)$$

If further $H_1(F) = 0$ then

$$H_3(X) \to H_3(Y) \to H_2(F)_{\pi_1} \to H_2(X) \to H_2(Y) \to 0 \qquad (1.8)$$

is exact. Further, there is the long exact sequence of homotopy groups π_i (for all $i \geq 0$);

$$\cdots \to \pi_{i+1}(Y) \to \pi_i(F) \to \pi_i(X) \to \pi_i(Y) \to \cdots . \qquad (1.9)$$

We have assumed $\pi_{i+1}(Y) = 0$ for $i \geq 1$ and $\pi_1(X) \to \pi_1(Y)$ is an isomorphism, so $\pi_2(F) \xrightarrow{\sim} \pi_2(X)$ ($\sim$ means isomorphism). and $\pi_1(F) = (0)$. The Hurewicz theorem now gives an isomorphism $\pi_2(F) \xrightarrow{\sim} H_2(F)$ and a commutative diagram

$$
\begin{array}{ccc}
\pi_2(F) & \xrightarrow{\;\sim\;} & H_2(F) \\
\sim\downarrow & & \downarrow \\
\pi_2(X) & \longrightarrow & H_2(X).
\end{array}
$$

Thus image of $H_2(F) \to H_2(X)(=$ image of $H_2(F)/\overline{\mathbb{R}\pi_1}H_2(F) \to H_2(X))$
$$= \text{image } \pi_2(X) \to H_2(X)$$
$$= (\text{definition}) : H_2^{\text{spherical}}(X) = H_2^{sph}(X).$$

Now (1.7) and (1.8) together become

$$H_3(X) \to H_3(Y) \to H_2(F)_{\pi_1} \to H_2(X) \to H_2(Y) \to 0 \to H_1(X) \xrightarrow{\sim} H_1(Y) \to 0.$$

So we get a short exact sequence

$$0 \to H_2^{sph}(X) \to H_2(X) \to H_2(Y) \to 0$$

$$(H_2(Y) = H_2(B\pi_1(X)))$$

due to Hopf.

Now we apply $\bar{\Delta}_* : H_2 \to H_1 \otimes H_1$, to the various spaces, and note that

$$\bar{\Delta}_* H_2(\mathbb{S}^2) = 0 \quad \text{as } H_1(\mathbb{S}^2) = 0.$$

So, $\bar{\Delta}_* H_2^{sph}(X) = 0$, giving a diagram

$$
\begin{array}{ccccccc}
0 & \longrightarrow & H_2^{sph}(X) & \longrightarrow & H_2(X) & \longrightarrow & H_2(Y) & \longrightarrow & 0 \\
& & \downarrow & & {\bar{\Delta}_*}\downarrow & & {\bar{\Delta}_*}\downarrow & & \\
& 0 & \longrightarrow & & H_1(X) \otimes H_1(X) & \xrightarrow{\sim} & H_1(Y) \otimes H_1(Y). & &
\end{array}
$$

So image $\bar{\Delta}_* \subset H_1(X) \otimes H_1(X)$ can be identified with image $\bar{\Delta}_* \subset H_1(Y) \otimes H_1(Y)$, $Y = B(\pi_1(X))$.

Next we look at spaces $Y = B(\text{group})$: for any normal subgroup N of a group Γ (discrete), we have a fibration

$$B(N) \to B(\Gamma) \to B(\Gamma/N).$$

We write $H_i(B(\text{group}); \mathbb{Z}) = H_i(\text{group}; \mathbb{Z})$. Then the exact sequence (1.7) becomes, with coefficients $\mathbb{Z}$,

$$H_2(\Gamma; \mathbb{Z}) \to H_2(\Gamma/N; \mathbb{Z}) \to H_0(\Gamma/N; H_1(N)) \to H_1(\Gamma) \to H_1(\Gamma/N) \to 0. \tag{1.10}$$

Here

$$
\begin{aligned}
H_0(\Gamma/N; H_1(N)) &= H_1(N)/\overline{\mathbb{Z}\Gamma} H_1(N) \\
&= (N/(N,N))/((\Gamma,N)/(\Gamma,(N,N))) \\
&= N/(\Gamma,N).
\end{aligned}
$$

Using this in (1.10), we get a sequence with exact rows and diagonals

$$0 \qquad\qquad 0$$

$$\searrow \qquad \uparrow$$

$$N/N \cap (\Gamma, \Gamma)$$

$$\uparrow \qquad \searrow$$

$$H_2(\Gamma) \to H_2(\Gamma/N) \to \quad N/(\Gamma, N) \quad \to \Gamma/(\Gamma, \Gamma) \to (\Gamma/N)/(\Gamma/N, \Gamma/N) \to 0$$

$$\|$$

$$\Gamma/N(\Gamma, \Gamma)$$

Now take $\Gamma = \pi_1(X) = \pi$, $N = (\pi, \pi)$, $\Gamma/N = \pi^{ab}$, $N/(\Gamma, N) = \pi_1^{(2)}/\pi_1^{(3)}$. The exact sequence (1.10), with coefficients in $\mathbb{R}$ (but $\mathbb{Q}$ would also work) gives the exact upper row of the following diagram, whose lower row is exact by definition:

$$
\begin{array}{ccccccc}
H_2(\pi_1) & \longrightarrow & H_2(\pi_1^{ab}) & \longrightarrow & (\pi_1^{(2)}/\pi_1^{(3)}) \otimes \mathbb{R} & \longrightarrow & 0 \\
\| & & \downarrow{\scriptstyle \bar{\Delta}_* \text{ (for } \pi_1^{ab})} & & & & \\
H_2(\pi_1) & \xrightarrow{\bar{\Delta}_*} & H_1(\pi_1) \otimes H_1(\pi_1)_{skew} & \longrightarrow & \dfrac{(H_1 \otimes H_1)_{skew}}{\bar{\Delta}_*(H_2(\pi_1))} & \longrightarrow & 0
\end{array}
$$

$$(1.11)$$

$(H_1(\pi_1) = H_1(X) = H_1(Y) = H_1)$.

Lemma 1.1

$$H_2(\pi_1^{ab}; \mathbb{R}) \xrightarrow{\bar{\Delta}_*} H_1(\pi_1^{ab}; \mathbb{R}) \otimes_{\mathbb{R}} H_1(\pi_1^{ab}; \mathbb{R})_{skew}$$

is an isomorphism.

Proof: We may assume π_1 is finitely generated, then pass to direct limit. So π_1^{ab} is also finitely generated. Now $\pi_1^{ab} = \mathbb{Z}^r \oplus \text{Torsion}$ so $H_2(\pi_1^{ab}; \mathbb{R}) \xrightarrow{\sim} H_2(\mathbb{Z}^r; \mathbb{R})$.

Since $B(\mathbb{Z}) = \mathbb{S}^1$, $B(\mathbb{Z}^r) = \mathbb{S}^1 \times \cdots \times \mathbb{S}^1 (r \text{ times}) = T^r$: r-dimensional torus. Thus, $H_i(T^r) = \Lambda^i(H_1(T^r))$. Passing to the dual vector space $H^i(T^r)$ we see that $\bar{\Delta}^* = $ cup product: $\Lambda^2(H^1(T^r)) \to H^2(T^r)$ is an isomorphism. $\square$

So in the diagram (1.11) preceding this lemma the vertical $\bar{\Delta}_*$ is an isomorphism, and we conclude that the last vertical arrow is also an isomorphism(induced by $\bar{\Delta}_*$ on $H_2(\pi_1^{ab})$)

$$(\pi_1^{(2)}/\pi_1^{(3)}) \otimes \mathbb{R} \to \Lambda^2(H_1)/\bar{\Delta}_*(H_2).$$

(H_i are $H_i(X;\mathbb{R})$, $\pi_1 = \pi_1(X)$.)

We now conclude: $(\pi_1^{(2)}/\pi_1^{(3)})\otimes\mathbb{R} = Gr(\pi_1)_{(2)}$ has the same $\mathbb{R}$-dimension as $\bigwedge^2(H_1(\pi))/\bar{\Delta}_* H_2(\pi) = \mathfrak{g}_2$.

1.8 The basic isomorphisms

As consequence of this calculation, we have:

Assuming $\pi_1(X)^{ab} = H_1(X)$ is finitely generated, the previously defined surjective linear map

$$II \circ Q : Gr(\pi_1)_{(2)} \to \mathfrak{g}_2$$

is an isomorphism. We previously knew that

$$II \circ Q : Gr(\pi_1)_{(1)} = \pi_1/(\pi_1,\pi_1) \to \mathfrak{g}_1 = H_1$$

is an isomorphism (natural isomorphism), and induces $II \circ Q$ in degree 2 (as deg 1 elements generate deg 2 by commutator). Finally, we recall that $\mathfrak{g}$ is the free Lie algebra $L(H_1)$ on H_1 modulo the relations $\bar{\Delta}_*(H_2)$ contained in $[H_1, H_1]$. So the "identity map" $H_1 \to (Gr\pi_1)_{(1)}$ extends to a Lie algebra homomorphism

$$\varphi : L(H_1) \to (Gr\pi_1)\otimes\mathbb{R} = Gr\,.$$

We also have the natural quotient (onto) homomorphism $\psi : L(H_1) \to \mathfrak{g}$, all of these fitting into a triangle of Lie algebra homomorphisms, which commutes since it commutes on generators=degree 1 elements:

$$
\begin{array}{ccc}
L(\ H_1) & \xrightarrow{\ \varphi\ } & Gr(\pi_1)\ \otimes\mathbb{R} = Gr \\
\psi \searrow & & \swarrow II\circ Q \\
& \mathfrak{g} &
\end{array}
\qquad .
$$

In degree 2, $II \circ Q$ is an isomorphism and so $\ker\psi_2 = \bar{\Delta}_* H_2 = \ker\varphi_2$. Since $\varphi(\bar{\Delta}_* H_2) = 0$, φ induces a Lie algebra homomorphism $\mathfrak{g} \to Gr(\pi_1)\otimes \mathbb{R}$, which is inverse to $II \circ Q$. Thus

Theorem 1.4 *$II \circ Q$, II, and Q are all Lie algebra isomorphisms (of the associated graded algebras):*

$$Q : (Gr\pi_1)\otimes\mathbb{R} \to P(Gr(\overline{\mathbb{R}\pi_1}))$$
$$II : P(Gr(\overline{\mathbb{R}\pi_1})) \to \mathfrak{g} = L(H_1)/(\bar{\Delta}_* H_2)$$

and so also isomorphisms of the universal enveloping algebras

$$Q : U((Gr\pi_1) \otimes \mathbb{R}) \to Gr(\mathbb{R}\pi_1) = U(P(Gr\mathbb{R}\pi))$$
$$II : Gr(\mathbb{R}\pi_1) \to U(\mathfrak{g}) = Gr(U(\mathfrak{g})^\wedge).$$

Finally, since II arises from a (Hopf) algebra homomorphism $\mathbb{R}\pi_1 \to U(\mathfrak{g})^\wedge$, and is an isomorphism on associated graded algebras, it is also an isomorphism on completions:

$$II : (\mathbb{R}\pi_1)^\wedge \to U(\mathfrak{g})^\wedge$$

is an isomorphism.

1.9 Lattices in nilpotent Lie groups

The term "lattice" here means a subgroup D of a (connected, simply-connected) nilpotent Lie group N such that D is discrete in N and N/D is compact.

Theorem 1.5 *Assume $\pi_1(X, x_0)$ is finitely generated. Let $G_{(n)} = \exp(\mathfrak{g}/\mathfrak{g}_{\geq n}) = \exp(\mathfrak{g}_{(n)})$ as before, and $II : \pi_1(X; x_0) \to G \to G_{(n)}$ the monodromy map given by iterated integration of Chen's connection θ. Then, the image of $\pi_1(X, x_0)$ in $G_{(n)}$ is discrete and co-compact; i.e., $G_{(n)}/(image)$ is a compact manifold, for each $n \geq 2$.*

Further, this image is just $\pi_1(X, x_0)$ modulo ($\pi_1^{(n)}$ and torsion elements mod $\pi_1^{(n)}$).

Proof: We use induction on n and the following lemma:

Let G be any finite dimensional connected Lie group and N a connected closed normal subgroup of G (which we may assume is also a Lie group), $G/N = H$ the quotient Lie group.

Suppose $D \subset G$ is a subgroup such that $D \cap N$ is a discrete co-compact subgroup of N, and $D/D \cap N \subset G/N$ is also discrete co-compact. Then $D \subset G$ is discrete co-compact.

Proof of Lemma We consider the principal fiber bundle $N \xrightarrow{i} G \xrightarrow{p} G/N$. Since $D/D \cap N$ is discrete in G/N, there is a neighborhood U of the identity eN in G/N such that $U \cap (DN/N) = eN$. If U is small enough, there is a section $s : U \rightarrow p^{-1}(U)$, $s(eN) = e$. Then $p^{-1}(U) = s(U)N$ is diffeomorphic to $U \times N$ (by multiplication) and $D \cap s(U)N = D \cap N$.

Since $D \cap N$ is discrete in N, there is a neighborhood V of e in N so that $D \cap V = (e)$. Now $s(U)V$ is a neighborhood of e in G and $D \cap s(U)V = (D \cap N) \cap V = D \cap V = (e)$, so D is discrete in N.

To see G/D is compact, note that G/N is the union of the $D/D \cap N$ translates of finitely many compact sets, and N is the union of the $D \cap N$ translates of finitely many compact sets. $\square$

Next we prove the theorem.

Recall our notation: $\mathfrak{g}_{(n)} = \mathfrak{g}/\mathfrak{g}_{\geq n}$, $G_{(n)} = \exp \mathfrak{g}_{(n)}$ Furthermore $\mathfrak{g}_n = [[\mathfrak{g}_1, \mathfrak{g}_1], \cdots, \mathfrak{g}_1]$ (n factors) denoted $\mathfrak{g}_1^{[n]}$. And so $\mathfrak{g}^{[n]} = \mathfrak{g}_{\geq n}$, and $\mathfrak{g}_{(n)} = \mathfrak{g}/\mathfrak{g}^{[n]}$, $G_{(n)} = G/G^{(n)}$ ($G^{(n)}$ denoting the iterated commutator subgroup).

We now have two exact sequences of groups related by the group homomorphisms $II : \pi \rightarrow G_{(n)}$.

$$(1) \longrightarrow G^{(n)}/G^{(n+1)} \longrightarrow G_{(n+1)} \longrightarrow G_{(n)} \longrightarrow (1)$$
$$\big\uparrow II \qquad\qquad \big\uparrow II \qquad\qquad \big\uparrow II$$
$$(1) \longrightarrow \pi^{(n)}/\pi^{(n+1)} \longrightarrow \pi/\pi^{(n+1)} \longrightarrow \pi/\pi^{(n)} \longrightarrow (1).$$

We need to prove that the leftmost vertical arrow is an isomorphism when we tensor $\pi^{(n)}/\pi^{(n+1)}$ with $\mathbb{R}$. To see this we show that the following diagram commutes (all maps are group homomorphisms as the groups and Lie algebras are abelian):

$$G^{(n)}/G^{(n+1)} \xrightarrow{\ \log\ } \mathfrak{g}^{[n]}/\mathfrak{g}^{[n+1]} = \mathfrak{g}_n$$
$$\big\uparrow II \qquad\qquad\qquad \big\uparrow GrII$$
$$\pi^{(n)}/\pi^{(n+1)} \xrightarrow{\ Q\ } Prim(Gr\mathbb{R}\pi)_n.$$

Here $GrII$ is the isomorphism $Gr\mathbb{R}\pi \rightarrow U(\mathfrak{g})$, $Prim$ denotes primitive elements (which map isomorphically to $\mathfrak{g}$).

For $\gamma \in \pi^{(n)}$, $II(\gamma) \in G^{(n)}$ since II is a group homomorphism, and $G^{(n)} = \exp \mathfrak{g}^{[n]}$, so

$$II(\gamma) = 1 + x_n, \qquad \mod U(\mathfrak{g}_{\geq n+1}^{\wedge})$$

with $x_n \in \mathfrak{g}_n$. Thus $\log II(\gamma) = x_n \mod \mathfrak{g}_{\geq n+1} = \mathfrak{g}^{[n+1]}$.

On the other hand,

$$Q(\gamma\pi^{(n+1)}) = (\gamma - 1) \qquad \mathrm{mod}\ (\overline{\mathbb{R}\pi})^{n+1},$$

and

$$GrII(Q(\gamma\pi^{(n+1)}) = II(\gamma - 1) = II(\gamma) - 1 \qquad \mathrm{mod}\ \mathfrak{g}^{[n+1]}$$
$$= x_n \text{ again.}$$

We conclude by recalling that $GrII$ is an isomorphism for each n. $\square$

1.10   Some Hodge theory

We have still to construct the differential forms $\mathcal{H}^1, C^1, \mathcal{H}^2, E^2$ used in the connection θ and for this purpose we will give an introduction to Hodge theory on Riemannian and Kahlerian manifolds, then define these 1- and 2-forms. This will complete the proofs of the discussion of π_1 and nilpotent Lie groups.

A. Compact Riemannian and Kahlerian Manifolds

1. Riemannian manifolds X.

Start with X any C^∞ manifold of dimension n, $A^p(X) = $ all C^∞ p-forms. We assume that for each $x \in X$ with $T_x X$ the tangent space, we are given a positive definite symmetric bilinear form $g_x = g : T_x X \times T_x X \to \mathbb{R}$.

g gives an isomorphism $G : T_x \to T_x^* = $ cotangent space, by $G(v)(w) = g(v, w)$ and so we can define g as a bilinear form $T^* \times T^* \to \mathbb{R}$, $g(G(v), G(w)) = g(v, w)$. We assume g is differentiable with respect to x.

For each integer p, $0 \le p \le n = dimX$ we can define $g : \wedge^p T^* \otimes \wedge^p T^* \to \mathbb{R}$, again positive definite symmetric, by

$$g(l_1 \wedge \cdots \wedge l_p, l'_1 \wedge \cdots \wedge l'_p) = \det[g(l_i, l'_j)].$$

Next we assume X is oriented, *i.e.*, we have a (continuous) choice of one of the two connected components of $(\wedge^n T_x^*) - (0)$, called the "orientation class".

Definition 1.2   The volume element $dvol \in \wedge^n T_x^*$ is the unique element in the orientation class satisfying $g(dvol, dvol) = 1$.

If $e^1, \cdots e^n$ is an orthonormal basis of T^* with $e^1 \wedge \cdots e^n$ in the orientation class, then this n-covector is $dvol$.

Next we can define the Hodge $\ast$-operator $\wedge^p T_x^* \rightarrow \wedge^{n-p} T_x^*$, a linear isomorphism. First we recall the interior product operator

$$i_l : \wedge^p T^* \rightarrow \wedge^{p-1} T^*, \quad \text{for } l \in T^*$$

given by

$$
\begin{aligned}
i_l(l_1 \wedge \cdots \wedge l_p) = \ & g(l, l_1) l_2 \wedge \cdots \wedge l_p \\
& - g(l, l_2) l_1 \wedge l_3 \wedge \cdots \wedge l_p \cdots \\
& + (-1)^{p-1} g(l, l_p) l_1 \wedge \cdots \wedge l_{p-1}.
\end{aligned}
$$

Now we define $\star$ as follows: for $l_1 \wedge \cdots \wedge l_p \in \bigwedge^P T^*$, let

$$\star(l_1 \wedge \cdots \wedge l_p) = i_{l_p} i_{l_{p-1}} \cdots i_{l_1}(dvol).$$

A main property of $\star$ is: for α, β both $\in \wedge^p T^*$,

$$\alpha \wedge \star \beta = g(\alpha, \beta) dvol = \beta \wedge \star \alpha.$$

Also, $\star : \wedge^p \rightarrow \wedge^{n-p}$, $\star : \wedge^{n-p} \rightarrow \wedge^p$ and $\star\star = (-1)^{p(n-p)}$ on $\wedge^p$.

We may let the point x vary in the above constructions which take place for each T_x^*, and obtain the operator $\star$ taking p-forms $A^p(X)$ to $n-p$ forms $A^{n-p}(X)$ (all forms are real valued). Thus, for $\alpha, \beta \in A^p(X)$, $g(\alpha, \beta)$ is a C^∞ function and $g(\alpha, \beta) dvol = \alpha \wedge \star \beta \in A^n(X)$. Also, $g(\alpha, \alpha) \geq 0$ and $g(\alpha, \alpha) = 0$ implies $\alpha = 0$.

Suppose further X is compact. Then each $A^p(X)$ is a pre-Hilbert space with inner product

$$(\alpha, \beta) = \int_X \alpha \wedge \star \beta = (\alpha, \beta)_X.$$

Next we bring in the exterior differential d, and define an operator

$$d^* = \pm \star d \star : A^{p+1} \rightarrow A^p$$

so that for $\alpha \in A^p(X)$, $\gamma \in A^{p+1}(X)$,

$$(d\alpha, \gamma)_X = (\alpha, d^*\gamma)_X.$$

The exact sign in $d^* = \pm \star d \star$ is determined by using

$$\star_{n-p}(\star_p \alpha^p) = (-1)^{p(n-p)} \alpha^p$$

and Stokes's theorem. For n even $d^* = - \star d \star$.

The Laplace operator $\Delta : A^p(X) \to A^p(X)$ is defined as

$$\Delta = dd^* + d^*d.$$

Definition 1.3 $\mathcal{H}^p = \ker \Delta = \{h \in A^p : dd^*h + d^*dh = 0\}.$

Equivalently, $\mathcal{H}^p = \{h \in A^p : dh = 0 \text{ and } d^*h = 0\}.$

It is easy to calculate that the three subspaces $\mathcal{H}^p, d^*A^{p+1} =$ image d^*, $dA^{p-1} =$ image d are mutually orthogonal. We will assume known (or accepted) the following basic theorem from elliptic PDE.

Theorem 1.6   *Let X be compact oriented Riemannian and let $\beta \in A^p$ be orthogonal to $\mathcal{H}^p$; that is, for all $h \in \mathcal{H}^p$, $\int_X \beta \wedge \star h = 0$. Then there is a unique $\alpha \in A^p$ such that*

$$\Delta\alpha = \beta.$$

Further, $\mathcal{H} = \ker \Delta$ is finite dimensional.

Assuming this theorem, we prove that A^p is the orthogonal direct sum of the three subspaces $\mathcal{H}^p, dA^{p-1}, d^*A^{p+1}$:

$$A^p = \mathcal{H}^p \oplus dA^{p-1} \oplus d^*A^{p+1} \quad \text{(Hodge decomposition)}.$$

To see this, start with any $\gamma \in A^p$ and let

$$\beta = \gamma - (\text{orthogonal projection of } \gamma \text{ on } \mathcal{H})$$
$$= \gamma - P(\gamma) = \gamma - \sum_i (\gamma, h_i)h_i$$

where h_i are an orthonormal basis of $\mathcal{H}^p$. By the theorem, there is a unique α so that

$$\beta = \gamma - P(\gamma) = \Delta\alpha = dd^*\alpha + d^*d\alpha,$$

so

$$\gamma = P(\gamma) + d(d^*\alpha) + d^*(d\alpha).$$

Next we show that $d : d^*A^{p+1} \to dA^p$ is an isomorphism, and $d^* : dA^p \to d^*A^{p+1}$ is an isomorphism. To show d is onto, we just note that $d(\mathcal{H}^p) = 0, d(dA^{p-1}) = 0$ so $dA^p = d(d^*A^{p+1})$. To show d is 1-1 on d^*A^{p+1}, let $\gamma \in d^*A^{p+1}$ and $d\gamma = 0$. Then $d^*\gamma = 0$ as $d^*d^* = 0$ $(d^*d^* = (dd)^*)$. Now $d\gamma = 0$, $d^*\gamma = 0$ implies $\gamma \in \mathcal{H}$, but $\mathcal{H}$ and d^*A^{p+1} are orthogonal, so $\gamma = 0$. Similarly d^* is an isomorphism.

We conclude that

$$\ker d = \mathcal{H}^p \oplus dA^{p-1}$$

and so $\mathcal{H}^p \to \ker d \to \ker d/\operatorname{Im}d = H^p_{DR}(X)$ is an isomorphism. In other words, the natural onto map $\ker d \to \ker d/\operatorname{Im}d = H^p$ is "split" by the choice of a complement to $\operatorname{Im}d$ in $\ker d$, namely the orthogonal complement $\mathcal{H}^p$. The Hodge $\star$ operator sends $\mathcal{H}^p$ isomorphically to $\mathcal{H}^{n-p}$, dA^{p-1} to $d^* A^{n-p+1}$ and $d^* A^{p+1}$ to dA^{n-p+1}.

Note that $\star$ does not behave simply with respect to wedge products or d, and in general the wedge product of two elements of $\mathcal{H}$ need not lie in $\mathcal{H}$: if $\alpha, \beta \in \mathcal{H}$ then $\alpha \wedge \beta = \gamma + d\eta$, $\gamma \in \mathcal{H}$, $d\eta \in dA$.

B. Complex manifolds, Kahler manifolds

First consider $\mathbb{C}^n$ with complex valued coordinate fuctions $z_j = x_j + iy_j$. By tangent space at a point p we mean the tangent space of the underlying real manifold with real coordinates $x_1, y_1, \cdots, x_n, y_n$. Thus, T_p has $\mathbb{R}$-basis

$$\frac{\partial}{\partial x_1}, \frac{\partial}{\partial y_1}, \cdots, \frac{\partial}{\partial x_n}, \frac{\partial}{\partial y_n}.$$

We define the ($\mathbb{R}$-linear) operator $J : T_p \to T_p$ by

$$J\left(\frac{\partial}{\partial x_i}\right) = \frac{\partial}{\partial y_i}, \quad J\left(\frac{\partial}{\partial y_i}\right) = -\frac{\partial}{\partial x_i}, \quad i = 1, \cdots n.$$

(so T_p is n-dimensional over the field $\mathbb{R} + \mathbb{R}J$ isomorphic to $\mathbb{C}$). On the dual vector space T_p^* with dual basis $dx_1, dy_1, \cdots, dx_n, dy_n$, the transpose of J, denoted J again, acts by

$$J(dy_i) = dx_i, \quad J(dx_i) = -dy_i.$$

Let U, V be open subsets of $\mathbb{C}^n$ and $F : U \to V$ a C^∞ map. We say that F is holomorphic if the differential $dF : T_p(U) \to T_{F(p)}(V)$ satisfies

$$dF \circ J = J \circ dF \quad (\text{all } p \in U).$$

These are the Cauchy-Riemann equations.

We can now define a complex manifold X to be a real $2n$ dimensional differentiable manifold with linear automorphisms J of $T_p(X)$ for each p, with $J^2 = -Id$, such that the coordinate charts F taking open sets U in $\mathbb{C}^n$ to open sets V in X satisfy $dF \circ J = J \circ dF$ (and so the "change of coordinates" $F_1^{-1} \circ F_2$, mapping an open set in $\mathbb{C}^n$ to another open set in $\mathbb{C}^n$, are holomorphic.)

Next we proceed to the definition of Kahler structure on a complex manifold X with J as above. First of all we assume given a Riemannian metric g on X such that if we restrict g to any T_p thern J preserve g (or J is orthogonal with respect to g), meaning

$$g(Jv, Jw) = g(v, w) \quad \text{for all } v, w \in T_P.$$

The standard example is $\mathbb{C}^n$ with the previous J and $\frac{\partial}{\partial x_1}, \cdots, \frac{\partial}{\partial y_n}$ being orthonormal with respect to g.

Next we want to construct a 2-form ω on X using g and J: we define

$$\omega(v, w) = -g(v, Jw) \quad \text{for } v, w \in T_p.$$

Equivalently, $\omega(v, Jw) = g(v, w)$. Then, $\omega(w, v) = -\omega(v, w)$ because

$$\begin{aligned} \omega(v, v) &= -g(v, Jv) = -g(Jv, J^2 v) \\ &= g(Jv, v) = g(v, Jv) = 0. \end{aligned}$$

Note that if $v \neq 0$, $g(v, v) > 0$ is equivalent to $\omega(v, Jv) > 0$: this says that ω restricted to the 2-dimensional space with ordered basis v, Jv is an orientation.

In $\mathbb{C}^n$ with Euclidean g as before, $\omega(\frac{\partial}{\partial x_i}, \frac{\partial}{\partial y_i}) = 1$ (and $\omega = 0$ on other pairs) says that $\omega = dx_1 \wedge dy_1 + \cdots + dx_n \wedge dy_n$.

Definition 1.4   A complex manifold (X, J) with Riemannian metric g such that J preserves g, is called a Kahler manifold (or has a "Kahler structure") if the 2-form $\omega(v, w) = -g(v, Jw)$ is a closed 2-form: $d\omega = 0$.

The simplest examples are $\mathbb{C}^n$ or open subsets U of $\mathbb{C}^n$ with Euclidean g, and quotients of $\mathbb{C}^n$ by a group of holomorphic isometries acting discontinuously with no fixed points. Every Kahler manifold is locally "like $\mathbb{C}^n$", where locally is to be interpreted in a suitable infinitesimal sense. For an explanation of this one may consult the book by Griffith and Harris, and we will just say that to prove certain identities involving d, J, d^*, on a Kahler manifold it is only necessary to check them in the standard case of $\mathbb{C}^n$. The identities we will need involve the following:

Define $d^c = (J^{-1} \circ d \circ J)/4\pi : A^p \to A^{p+1}$ (so

$$d^c = i(\bar{\partial} - \partial)/4\pi$$

and

$$dd^c = i\partial\bar{\partial}/2\pi).$$

Note that d^c as well as d takes real valued forms to real valued forms.

Define $(d^c)^* =$ adjoint of d^c (relative to $(\alpha, \beta)_X = \int_X \alpha \wedge \star\beta$) so

$$(d^c)^* = J^{-1}d^*J/4\pi \quad (\text{as } J^* = J^{-1}).$$

The identities referred to above are:

$$dd^c = -d^c d,$$
$$dd^{c*} = -d^{c*}d,$$
$$d^c d^* = -d^* d^c$$

(however $dd^* + d^*d = \Delta \neq 0$, $d^c d^{c*} + d^{c*}d^c = \Delta/16\pi^2 \neq 0$).

J is defined to act on forms so that $J(\alpha \wedge \beta) = J\alpha \wedge J\beta$. Also

$$J \circ \star = \star \circ J.$$

As an easy consequence of these identities, *i.e.*, of the fact that d and d^* each commutes (up to a - sign) with d^c and d^{c*}, we have a further decomposition of each $A^p(X)$ into orthogonal subspaces:

$$A^p(X) = \mathcal{H}^p \oplus dd^c A^{p-2} \oplus dd^{c*} A^p \oplus d^* d^c A^p \oplus d^* d^{c*} A^{p+2}$$

(here $d^{c*} = J^{-1}d^*J/4\pi$). Further, dd^c maps $d^* d^{c*} A^{p+2}$ isomorphically onto $dd^c A^p \subset A^{p+2}$ and is 0 on the other subspaces, and similarly each of the operators $dd^{c*}, d^c d^*, d^* d^{c*}$ is an isomorphism on one of the above subspaces and is 0 on the others. This implies the important

Lemma 1.2 $(d, d^c$ *lemma*) *Suppose* $\alpha \in A^p$ *is d exact and d^c closed, or else α is d^c exact and d closed. Then α is dd^c exact.*

Proof: We note that

$$\mathrm{Im}\, d = dd^c A^{p-2} \oplus dd^{c*} A^p$$
$$\ker d^c = \mathcal{H}^p \oplus dd^c A^{p-2} \oplus d^* d^c A^p,$$

so $\mathrm{Im}\, d \cap \ker d^c = dd^c A^{p-2}$. Similarly $\mathrm{Im}\, d^c \cap \ker d = dd^c A^{p-2}$. Another simple but important fact is:

$$\ker d \cap \ker d^c = \mathcal{H}^p \oplus dd^c A^{p-2}$$

so that $(\ker d \cap \ker d^c)/\mathrm{Im}\, dd^c$ is naturally isomorphic both to $\mathcal{H}^p$ and to $H^p_{DR}(X)$. The isomorphism of this quotient vector space with H^p_{DR} involves only the complex structure J of X and not the choice of a Kähler metric g (but assumes a Kähler metric exists).

We can now complete the proof of Chen's theorem in the Kahler case by exhibiting the subspaces $\mathcal{H}^p, C^p, E^p$ satisfying the condition 1., 2., 3a., 3b needed.

We begin by considering the subalgebra $\ker d^c$ of A^p for $p = 1, 2$, so $\ker d^c = \mathcal{H}^p \oplus d^c dA^{p-2} \oplus d^c d^* A^p$. The subspace $\mathcal{H}^p$ of Chen's theorem is taken to be just the harmonic p-forms $\mathcal{H}^p$, the subspace C^1 is defined to be $d^c d^* A^p$, for $p = 1$, and the subspace E^2 is defined to be $d^c dA^{p-2}$, for $p = 2$, i.e., $d^c dA^0$. Thus $d : C^1 \to E^2$ is an isomorphism. As noted previously, the conditions 1., 2., 3a are valid in a Riemannian manifold, if we take $\mathcal{H}, C, E$ to mean: harmonic, coexact , exact.

Condition 3b was

$$[(\mathcal{H}^1 + C^1) \wedge C^1] \cap \ker d \subset E^2.$$

To see this, we write

$$(\mathcal{H} \oplus d^c d^* A) \wedge d^c d^* A \subset d^c A$$

(since d^c is a super-derivation and $d^c \mathcal{H} = 0$), and

$$d^c A = d^c dA \oplus d^c d^* A,$$

so $d^c A \cap \ker d = d^c dA = E$. $\square$

Chapter 2

Iterated Integrals on Compact Riemann Surfaces

2.1 Introduction

Our next main purpose is to study Chen's connection θ more closely in the lowest-dimensional case, namely when X is a compact Riemann surface with complex structure J and (Kahler) metric g (note that the Kahler condition $d\omega = 0$ is automatic here since X has real dimension 2). To do this we will make many simplifying assumptions on the terms of θ on which we will concentrate, in particular we construct a simplified version $\bar{\theta}$ of θ; more precisely the series defining the homomorphism of $\pi_1(X, x_0)$ given by $\bar{\theta}$ will contain some of the terms of the series given by θ. A main reference will be [Harris, 1983a].

2.2 Generalities on Riemann surfaces and iterated integrals

Note that the $\star$ operator on 1-forms of X is conformally invariant, so depends only on the complex structure and not on the choice of metric in its conformal equivalence class (this is always true in the middle dimension), in fact $\star = J^{-1}$ on 1-forms here. Thus,

$$\mathcal{H}^1 = \ker d \cap \ker d^* = \ker d \cap \ker d^c,$$
$$C^1 = d^c d^* A^1 = d^* d A^1 = d^* A^2 = d^c A^0,$$
$$E^2 = dC^1 = dd^c A^0,$$

so all these subspaces are definable by d, d^c alone, when X is a Riemann surface. Now let $\alpha_1, \alpha_2 \in \mathcal{H}^1$ satisfy

$$\int_X \alpha_1 \wedge \alpha_2 = 0;$$

35

equivalently, $\alpha_1 \wedge \alpha_2$ is exact, so

$$\alpha_1 \wedge \alpha_2 + d\alpha_{12} = 0$$

for a unique $\alpha_{12} \in C^1 = d^c A^0$.

Let now A be the associative algebra with 2 generators x, y and relations

$$x^2 = y^2 = xyx = yxy = 0.$$

A has $\mathbb{R}$-basis $1, x, y, xy, yx$ and contains the Lie algebra $\bar{L} = L/L^{(3)}$ with basis $x, y, [x, y]$ (L=free Lie algebra on generators x, y). Let $\bar{\theta}$ be the $\bar{L}$ valued 1-form

$$\bar{\theta} = \alpha_1 \otimes x + \alpha_2 \otimes y + \alpha_{12} \otimes [x, y].$$

Then

$$d\bar{\theta} + \frac{1}{2}[\bar{\theta}, \bar{\theta}] = 0$$

and so $\bar{\theta}$ defines a homomorphism

$$t : \pi_1(X, x_0) \to G = \exp(\bar{L})$$
$$= \text{elements of the form } \exp(ax + by + c[x, y]) \text{ in } A.$$

t is in fact a homomorphism of the fundamental groupoid (not necessarily closed paths, modulo end-point preserving homotopy) into G. This implies

(1) $\int_{\gamma_1 \gamma_2} \alpha_i = \int_{\gamma_1} \alpha_i + \int_{\gamma_2} \alpha_i$,

(2) $\int_{\gamma_1 \gamma_2} (\alpha_1, \alpha_2) + \alpha_{12}$

$$= \int_{\gamma_1} (\alpha_1, \alpha_2) + \alpha_{12} + \int_{\gamma_2} (\alpha_1, \alpha_2) + \alpha_{12} + \int_{\gamma_1} \alpha_1 \int_{\gamma_2} \alpha_2,$$

(3) $\int_{\gamma^{-1}} (\alpha_1, \alpha_2) + \alpha_{12} = -\int_{\gamma} (\alpha_1, \alpha_2) + \alpha_{12} + \int_{\gamma} \alpha_1 \int_{\gamma} \alpha_2.$

Using (2), (3) we derive the following formula for changing base point in π_1: let x_0, x_1 be base points, l a path x_1 to x_0 (up to homotopy) $\gamma \in \pi_1(X, x_0)$, $l\gamma l^{-1} \in \pi_1(X, x_1)$, then

(4) $\int_{l\gamma l^{-1}} (\alpha_1, \alpha_2) + \alpha_{12} = \int_{\gamma} (\alpha_1, \alpha_2) + \alpha_{12} + \int_l \alpha_1 \int_{\gamma} \alpha_2 - \int_{\gamma} \alpha_1 \int_l \alpha_2.$

Our aim is to simplify the algebraic situation by making further assumptions on $\alpha_1, \alpha_2, \gamma$. First, we assume α_1, α_2 satisfy: $\int_{\gamma} \alpha_i \in \mathbb{Z}$ for all $\gamma \in \pi_1(X, x_0)$. We denote this by $\alpha_i \in \mathcal{H}_{\mathbb{Z}}^1$. We retain the assumption

$\int_X \alpha_1 \wedge \alpha_2 = 0$ and introduce a subgroup $(\mathcal{H}_{\mathbb{Z}}^1 \otimes \mathcal{H}_{\mathbb{Z}}^1)'$ of $\mathcal{H}_{\mathbb{Z}}^1 \otimes \mathcal{H}_{\mathbb{Z}}^1$, namely the kernel of

$$\mathcal{H}_{\mathbb{Z}}^1 \otimes \mathcal{H}_{\mathbb{Z}}^1 \to \mathbb{Z}, \quad \alpha \otimes \beta \mapsto \int_X \alpha \wedge \beta.$$

Thus $\alpha_1 \otimes \alpha_2 \in (\mathcal{H}_{\mathbb{Z}}^1 \otimes \mathcal{H}_{\mathbb{Z}}^1)'$ and as before, α_{12} satisfies

$$\alpha_1 \wedge \alpha_2 + d\alpha_{12} = 0.$$

Such a form α_{12} is defined for any element of $(\mathcal{H}_{\mathbb{Z}}^1 \otimes \mathcal{H}_{\mathbb{Z}}^1)'$. We now note that if we write

$$I(\alpha_1, \alpha_2; \gamma) = \int_\gamma (\alpha_1, \alpha_2) + \alpha_{12}, \quad \mod \mathbb{Z}, \quad \text{in } \mathbb{R}/\mathbb{Z},$$

then by (2), I defines a function of 2 variables

$$I : (\mathcal{H}_{\mathbb{Z}}^1 \otimes \mathcal{H}_{\mathbb{Z}}^1)' \times \pi_1(X, x_0) \to \mathbb{R}/\mathbb{Z}$$

which is "bi-multiplicative", and since $\mathbb{R}/\mathbb{Z}$ is an abelian group, factors through a homomorphism still denoted I, or I_{x_0},

$$I_{x_0} : (\mathcal{H}_{\mathbb{Z}}^1 \otimes \mathcal{H}_{\mathbb{Z}}^1)' \otimes \pi_1(X, x_0)^{ab} \to \mathbb{R}/\mathbb{Z}.$$

This homomorphism depends on the base point x_0, as shown by formula (4). In order to remove the dependence on x_0, we restrict I_{x_0} to a smaller subgroup: namely the *kernel* of the homomorphism

$$(\mathcal{H}_{\mathbb{Z}}^1 \otimes \mathcal{H}_{\mathbb{Z}}^1)' \otimes \pi_1(X, x_0)^{ab} \to \mathcal{H}_{\mathbb{Z}}^1 \oplus \mathcal{H}_{\mathbb{Z}}^1$$
$$\sum \alpha_{1i} \otimes \alpha_{2i} \otimes [\gamma_i] \mapsto \sum (\int_{\gamma_i} \alpha_{2i})\alpha_{1i} \oplus \sum -(\int_{\gamma_i} \alpha_{1i})\alpha_{2i}.$$

(4) then shows that the definition of I on the subgroup is unchanged if we replace each γ_i by $l\gamma_i l^{-1}$. We can then replace $\pi_1(X, x_0)$ by $H_1(X; \mathbb{Z})$ and use the Poincaré duality isomorphism $H_1(X; \mathbb{Z}) \to \mathcal{H}_{\mathbb{Z}}^1$ (which we define so that $[\gamma] \mapsto \alpha_\gamma = \alpha_3$ implies

$$\int_\gamma \alpha = \int_X \alpha \wedge \alpha_\gamma$$

for all $\alpha \in \mathcal{H}^1$). Finally we denote the domain of I (the kernel just defined) as $(\mathcal{H}_{\mathbb{Z}}^1 \otimes \mathcal{H}_{\mathbb{Z}}^1 \otimes \mathcal{H}_{\mathbb{Z}}^1)'$.

This group is the kernel of

$$\mathcal{H}^1_{\mathbb{Z}} \otimes \mathcal{H}^1_{\mathbb{Z}} \otimes \mathcal{H}^1_{\mathbb{Z}} \to \mathcal{H}^1_{\mathbb{Z}} \oplus \mathcal{H}^1_{\mathbb{Z}} \oplus \mathcal{H}^1_{\mathbb{Z}}$$

$$\alpha \otimes \beta \otimes \gamma \mapsto (\int_X \alpha \wedge \beta)\gamma \oplus \int_X (\beta \wedge \gamma)\alpha \oplus (\int_X \gamma \wedge \alpha)\beta.$$

We note that if $\alpha_1, \alpha_2, \alpha_3 \in \mathcal{H}^1_{\mathbb{Z}}$ satisfy $\int_X \alpha_i \wedge \alpha_j = 0$ for $i, j = 1, 2, 3$ then $\alpha_1 \otimes \alpha_2 \otimes \alpha_3 \in (\mathcal{H}^1_{\mathbb{Z}} \otimes \mathcal{H}^1_{\mathbb{Z}} \otimes \mathcal{H}^1_{\mathbb{Z}})'$ and

$$I(\alpha_1 \otimes \alpha_2 \otimes \alpha_3) = \int_\gamma (\alpha_1, \alpha_2) + \alpha_{12}, \quad \mod \mathbb{Z}$$

(where $\gamma \in \pi_1(X, x_0)$ corresponds to α_3).

Next we would like to replace the domain of I by a subgroup $(\Lambda^3 \mathcal{H}^1_{\mathbb{Z}})'$ of the third exterior power $\Lambda^3(\mathcal{H}^1_{\mathbb{Z}})$, defined as the kernel of a homomorphism to $\mathcal{H}^1_{\mathbb{Z}}$.

Thus consider the commutative diagram

$$
\begin{array}{ccccccccc}
0 & \longrightarrow & (\mathcal{H}^{1\,\otimes 3}_{\mathbb{Z}})' & \longrightarrow & (\mathcal{H}^1_{\mathbb{Z}})^{\otimes 3} & \longrightarrow & \mathcal{H}^1_{\mathbb{Z}} \oplus \mathcal{H}^1_{\mathbb{Z}} \oplus \mathcal{H}^1_{\mathbb{Z}} & \longrightarrow & 0 \\
& & \downarrow{\scriptstyle j_1} & & \downarrow{\scriptstyle j_2} & & \downarrow{\scriptstyle j_3} & & \\
0 & \longrightarrow & (\Lambda^3 \mathcal{H}^1_{\mathbb{Z}})' & \longrightarrow & \Lambda^3 \mathcal{H}^1_{\mathbb{Z}} & \overset{k}{\longrightarrow} & \mathcal{H}^1_{\mathbb{Z}} & \longrightarrow & 0
\end{array}
$$

Here

$$k(\alpha \wedge \beta \wedge \gamma) = (\int_X \alpha \wedge \beta)\gamma + (\int_X \beta \wedge \gamma)\alpha + (\int_X \gamma \wedge \alpha)\beta,$$

and $j_3(\alpha \oplus \beta \oplus \gamma) = \alpha + \beta + \gamma$. j_2 is the usual homomorphism and induces j_1.

We will show that $2I$ can be defined as a homomorphism $(\Lambda^3 \mathcal{H}^1_{\mathbb{Z}})' \to \mathbb{R}/\mathbb{Z}$ such that for triples $\alpha_1, \alpha_2, \alpha_3 \in \mathcal{H}^1_{\mathbb{Z}}$ satisfying the special condition that for $i = 1, 2, 3$, α_i is Poincaré dual to a simple closed curve C_i, and C_1, C_2, C_3 are mutually disjoint, then

$$2I(\alpha_1 \wedge \alpha_2 \wedge \alpha_3) = 2[\int_{C_3} (\alpha_1, \alpha_2) + \alpha_{12}].$$

Lemma 2.1   *Let C_i, $i = 1, 2, 3$ be disjoint oriented simple closed curves on X. Then their harmonic Poincaré dual forms α_i satisfy $\alpha_i \otimes \alpha_j \otimes \alpha_k \in (\mathcal{H}^{1\,\otimes 3}_{\mathbb{Z}})'$ and $I(\alpha_1 \otimes \alpha_2 \otimes \alpha_3)$ is invariant under cyclic permutations and is 0 (in $\mathbb{R}/\mathbb{Z}$) if any two of the α's are equal.*

Proof: Consider the surface X_i with two boundary components C'_i, C''_i obtained by cutting X along C_i. More precisely, the boundary ∂X_i is the disjoint union of C_i' and $-C_i''$, and to a point p on C_i correspond two points p' on C_i', p'' on C_i''.

Now integration of α_i on X_i from a base point x_0 to a variable point x defines a single-valued (harmonic) function h_i such that

a) $h_i(p'') - h_i(p') = 1$, and

b) $dh_i = \alpha_i$, (the result of integration is independent of path x_0 to x on X_i).

Let η be any 1-form on X (continuous). By a) we have

c)

$$\int_{C_i} \eta = \int_{C_i''} h_i\eta - \int_{C_i'} h_i\eta$$

$$= -\int_{\partial X_i} h_i\eta = -\int_{X_i} d(h_i\eta)$$

$$= -\int_{X_i} (\alpha_i \wedge \eta + h_i d\eta).$$

Let $\eta = -\alpha_{j,k}$, which is coexact and so orthogonal to the harmonic form $\star\alpha_i$; then $\int_{X_i} \alpha_i \wedge \alpha_{j,k} = 0$ and, since $d\alpha_{j,k} = -\alpha_j \wedge \alpha_k$.

d) $\int_{C_i} -\alpha_{jk} = -\int_{X_i} h_i\alpha_j \wedge \alpha_k$.

Next, let $X_{k,i}$ denote X cut along both C_k and C_i. Then $h_i dh_j + \alpha_{ij}$ is a closed form on $X_{k,i}$ and so

$$\int_{X_{k,i}} dh_k \wedge (h_i dh_j) = \int_{X_{k,i}} dh_k \wedge (h_i dh_j + \alpha_{ij}) = \int_{X_{k,i}} d(h_k h_i dh_j + h_k \alpha_{ij})$$

$$= \int_{C_k'-C_k''+C_i'-C_i''} (h_k h_i dh_j + h_k \alpha_{ij})$$

$$= -\int_{C_k} h_i dh_j - \int_{C_i} h_k dh_j - \int_{C_k} \alpha_{ij}$$

$$= -\int_{C_k} h_i dh_j - \int_{C_i} h_k dh_j - \int_{C_k} \alpha_{ij}.$$

Rearranging terms, we get

$$-\int_{X_{k,i}} h_i dh_j \wedge dh_k + \int_{C_i} h_k dh_j = -\int_{C_k} h_i dh_j - \int_{C_k} \alpha_{ij}.$$

Using d) and $\int_{C_i} h_k dh_j = -\int_{C_i} h_j dh_k$ we get

$$\int_{C_i} (h_j dh_k + \alpha_{jk}) = \int_{C_k} (h_i dh_j + \alpha_{ij})$$

which is cyclic invariance. Also

$$\int_{C_i} h_k dh_j + \alpha_{jk} = -\int_{C_i} (h_j dh_k + \alpha_{kj})$$

is clear, so we have skew symmetry of $I(\alpha_i \otimes \alpha_j \otimes \alpha_k)$ under all transpositions, but furthermore we have vanishing of this iterated integral if $\alpha_j = \alpha_k$ since

$$\int_{C_i} h_j dh_j = \frac{1}{2} \int_{C_i} dh_j \int_{C_i} dh_j = 0.$$

Thus we have vanishing also if any two of the α are equal, concluding the proof. $\square$

Lemma 2.2   *Consider triples of disjoint simple closed curves* C_1, C_2, C_3 *as above and write* $P = \Lambda^3(\mathcal{H}_{\mathbb{Z}}^1)'$, $C_1 \wedge C_2 \wedge C_3 \in P$ *instead of* $\alpha_1 \wedge \alpha_2 \wedge \alpha_3 \in P$. *Such "decomposable elements" generate* $\Lambda^3(\mathcal{H}_{\mathbb{Z}}^1)' = P$.

Proof: Let $A_1, B_1, \cdots, A_g, B_g$ be standard simple closed curves in X, where we think of A_i as going around the i-th hole, and B_i going around the corresponding i-th handle with intersection number $A_i \cdot B_i = 1$ and all other pairs being disjoint.

Consider three types of elements $C_i \wedge C_j \wedge C_k$:

a) $1 \leq i < j < k \leq g$ and each $C_l = A_l$ or B_l. It is clear that such elements are as described above, belong to P, are linearly independent over $\mathbb{Z}$, and form a basis of a direct summand of $\Lambda^3(\mathcal{H}_{\mathbb{Z}}^1)$, and so a direct summand of P. There are $2^3 \binom{g}{3}$ of them.

b) $(A_i + A_1) \wedge (B_i - B_1) \wedge C_k$ where i, k are distinct and > 1, and $C_k = A_k$ or B_k. The curves corresponding to $A_i + A_1$, $B_i - B_1$ can be taken as, respectively, surrounding both the first and i-th hole, or going around a handle separating these two holes (imagined as next to one another). There are $2(g-1)(g-2)$ such elements.

c) $(A_i + A_2) \wedge (B_i - B_2) \wedge C_1$. Here $C_1 = A_1$ or B_1 and $i > 2$. These are of the same type as b) and there are $2(g-2)$ of them.

Modulo type a) elements we may replace a type b), $(A_i + A_1) \wedge (B_i - B_1) \wedge C_k$ by $(A_i \wedge B_i - A_1 \wedge B_1) \wedge C_k$ and similarly replace type c) by $(A_i \wedge B_i - A_2 \wedge B_2) \wedge C_1$ (where $i > 2$).

These new elements just defined are still in P and belong to the subspace of Λ^3 spanned by decomposables $C_i \wedge C_j \wedge C_k$ where each C_l is A_l or B_l and two of the indices i, j, k are equal. Thus they belong to a direct summand complementary to that of the type a) elements, which involve 3 distinct indices. They are easily seen to be linearly independent and span a further direct summand. Finally, the total number of elemnets is $\binom{2g}{3} - 2g$ which is the dimension of P, concluding the proof of this lemma. $\square$

Corollary 2.1 *Let D be the subgroup of $[(\mathcal{H}_\mathbb{Z}^1)^{\otimes 3})]'$ generated by all elements of type corresponding to $C_1 \otimes C_2 \otimes C_3$ where the C_i are disjoint simple closed curves in X.*

Then

$$j_1 : (\mathcal{H}_\mathbb{Z}^{1\,\otimes 3})' \to \Lambda^3(\mathcal{H}_\mathbb{Z}^1)' = P$$

maps D onto P and $2I$ (twice the iterated integral homomorphism I : $(\mathcal{H}_\mathbb{Z}^{1\,\otimes 3})' \to \mathbb{R}/\mathbb{Z}$) when restricted to D factors through $j_1 : D \to P$ and defines a homomorphism denoted

$$\nu = 2I : P \to \mathbb{R}/\mathbb{Z}.$$

Proof: Clearly D is invariant under the permutation group action on $(\mathcal{H}_\mathbb{Z}^1)^{\otimes 3}$ and on $((\mathcal{H}_\mathbb{Z}^1)^{\otimes 3})'$. We have shown that for any $d \in D$ and σ any transposition, $I(d + \sigma(d)) = 0$.

The kernel of $j_3 : D \to P$ is the intersection $D \cap K$ where K is the subgroup of $(\mathcal{H}_\mathbb{Z}^1)^{\otimes 3}$ generated by all "monomials" $m = x \otimes y \otimes z$ with two of the factors x, y, z equal.

Let S be the subgroup of $(\mathcal{H}_\mathbb{Z}^1)^{\otimes 3}$ generated by all elements of the form $t + \sigma(t)$ where σ is a transposition. Then $2K \subset S$.

If $d \in D \cap K$, $d = k \in K$ then $2d = 2k \in S$ and

$$2I(d) = I(2d) = 0$$

since $2d \in S \cap D$ and $I = 0$ on $S \cap D$. $\square$

We can now look at the situation with Chen's connection θ and its homomorphic image $\bar\theta$ and compare it with the usual Abel-Jacobi map . Recall that $\bar\theta$ induced a homomorphism

$$II : \pi_1(X, x_0) \to G = \exp(\mathfrak{g})$$

where G is a simply-connected 3-dimensional nilpotent Lie group (over $\mathbb{R}$) and $\mathfrak{g}$ is a Lie algebra with generators x, y, with $[x, y] \neq 0$ but all further commutators 0.

$$\bar\theta = \alpha_1 \otimes x + \alpha_2 \otimes y + \alpha_{12} \otimes [x, y]$$

$(\alpha_1 \wedge \alpha_2 + d\alpha_{12} = 0)$.

For $\gamma \in \pi_1(X, x_0)$ such that $\int_\gamma \alpha_i = 0$, $i = 1, 2$ we have in $G = \exp(\mathfrak{g})$:

$$II(\gamma) = 1 + \int_\gamma \theta + \int_\gamma (\theta, \theta)$$

$$= 1 + (\int_\gamma \alpha_{12} + (\alpha_1, \alpha_2))[x, y]$$

$$\in \exp[\mathfrak{g}, \mathfrak{g}] = (G, G).$$

The usual Abel-Jacobi map is just the "first stage" of Chen's construction, however we will apply it to a different connection

$$\psi = \alpha_1 \otimes h_1 + \alpha_2 \otimes h_2 + \alpha_3 \otimes h_3$$

with values in a 3 dimensional commutative Lie algebra $\subset H_1(X; \mathbb{R})$, $h_i \in H_1$, $\int_{h_j} \alpha_i = \delta_{ij}$, where $\alpha_i \in \mathcal{H}^1_{\mathbb{Z}}(X)$ and $\alpha_1 \wedge \alpha_2 \wedge \alpha_3 \in P$.

We then obtain a map, the Abel-Jacobi map $A : X \to \mathbb{R}^3/\mathbb{Z}^3$ if we choose a base point $x_0 \in X$, namely

$$A(x) = (\int_{x_0}^x \alpha_1, \int_{x_0}^x \alpha_2, \int_{x_0}^x \alpha_3) \qquad \mod \mathbb{Z}^3.$$

We write $\mathbb{R}^3/\mathbb{Z}^3$ for the Lie group $\exp(\mathfrak{g})$ ($\mathfrak{g}=$ Abelian Lie algebra spanned by homology classes C_1, C_2, C_3 Poincaré dual to the α_i) modulo the image of π_1. Using this map A, we see that if we pull back to X, by A^*, the standard 1-forms $\xi_i = dx_i$ on $\mathbb{R}^3/\mathbb{Z}^3$ we get $A^*(dx_i) = \alpha_i$.

Thus if we assume $\int_X \alpha_i \wedge \alpha_j = 0$, $i, j = 1, 2, 3$, then the cycle $A(X) \subset \mathbb{R}^3/\mathbb{Z}^3$ is homologous to 0 in $\mathbb{R}^3/\mathbb{Z}^3$:

$$A(X) = \partial D$$

for some(non-unique) 3-chain D in $\mathbb{R}^3/\mathbb{Z}^3$.

2.3 Harmonic volumes and iterated integrals

We can now compare the two maps $X \to G/lattice$ using $\bar{\theta}$ and $A : X \to \mathbb{R}^3/\mathbb{Z}^3$ as follows:

Definition 2.1 The (co)volume or "harmonic volume" corresponding to a triple $\alpha_1, \alpha_2, \alpha_3 \in \mathcal{H}^1_{\mathbb{Z}}(X)$ satisfying $\int_X \alpha_i \wedge \alpha_j = 0$ for $i \neq j$ is the "volume" mod $\mathbb{Z}$ of any 3-chain D in $\mathbb{T}^3 = \mathbb{R}^3/\mathbb{Z}^3$ satisfying $\partial D = A(X)$

as 2-chains, *i.e.*,

$$\int_D \xi_1 \wedge \xi_2 \wedge \xi_3 \qquad \text{mod } \mathbb{Z}$$

where $A^*(\xi) = \alpha_i$ and $\xi_1 \wedge \xi_2 \wedge \xi_3$, regarded as cohomology class, generates $H^3(\mathbb{T}^3; \mathbb{Z})$. We orient $\mathbb{T}^3$ using $\xi_1 \wedge \xi_2 \wedge \xi_3$, $\xi_i = dx_i$ on $\mathbb{R}/\mathbb{Z}$.

Theorem 2.1 *With the above notation, and assuming the α_i Poincaré dual to disjoint simple closed curves C_i as before,*

$$\int_D \xi_1 \wedge \xi_2 \wedge \xi_3 = \int_{C_3} (\alpha_1, \alpha_2) + \alpha_{12} \qquad \text{mod } \mathbb{Z}$$

so harmonic volume equals iterated integral.

Proof: Cut X along C_3 as before, obtaining X_3 with boundary $C_3' - C_3''$. Let h_3 be the $\mathbb{R}$-valued function on X_3 defined by: $h_3(x) = \int_{x_0}^x \alpha_3$. Recall that

$$\int_{C_3} \alpha_{12} = \int_{X_3} h_3 \alpha_1 \wedge \alpha_2.$$

For $i = 1, 2$ we define $h_i : X \to \mathbb{R}/\mathbb{Z}$ again by the formula $h_i(x) = \int_{x_0}^x \alpha_i$. So h_i are also defined on X_3. Denote

$$\vec{h} : X_3 \to \mathbb{R}^2/\mathbb{Z}^2 \times \mathbb{R} = \mathbb{T}^2 \times \mathbb{R}$$

the map given by (h_1, h_2, h_3). So $\vec{h}$ on C_3', C_3'' are related by

$$\vec{h}(p'') = \vec{h}(p') + (0, 0, 1)$$

. $\vec{h}$ restricted to C_3' is null-homologous since

$$\int_{C_3'} \vec{H}^*(\xi_i) = \int_{C_3'} \alpha_i = 0 \quad \text{for} i = 1, 2.$$

Since π_1 and H_1 for $\mathbb{T}^2 \times \mathbb{R}$ are isomorphic $\vec{h}$ extends from $C_3' \to \mathbb{T}^2 \times \mathbb{R}$ to a 2-disk B' with boundary C_3':

$$\text{extended } \vec{h} : X_3 \cup B' \to \mathbb{T}^2 \times \mathbb{R}.$$

Since we have a homeomorphism $C_3'' \to C_3'$ we can extend $\vec{h}$ from C_3'' to a copty B'' of B' by:

$$\vec{h}(b'') = \vec{h}(b') + (0, 0, 1)$$

for corresponding points b', b'' in B', B''. We have now filled in the two holes C_3', C_3'' in X_3 with two disks B', B'' and extended the map $\vec{h}$ to the filled-in surface $X^* = X_3^* = X_3 \cup B' \cup B''$ where $\partial X_3 = C_3' - C_3''$, $\partial B' = C_3'$, $\partial B'' = C_3''$,

$$\text{as cycle } X^* = X_3^* = X_3 - B' + B''.$$

We claim now that $\vec{h}(X^*)$ bounds in $\mathbb{T}^2 \times \mathbb{R}$: we just have to check that $\int_{X^*} \vec{h}^*(\xi_1 \wedge \xi_2) = 0$. But,

$$\int_{X^*} \vec{h}^*(\xi_1 \wedge \xi_2) = \int_{X_3} \alpha_1 \wedge \alpha_2 - \int_{\vec{h}(B')} \xi_1 \wedge \xi_2 + \int_{\vec{h}(B'')} \xi_1 \wedge \xi_2.$$

The first term equals $\int_X \alpha_1 \wedge \alpha_2 = 0$. The last two terms add up to zero because $\vec{h}(B'') = \vec{h}(B') + (0,0,1)$ and $\xi_1 \wedge \xi_2$ is unchanged by this traslation.

Now consider the natural covering map

$$p : \mathbb{T}^2 \times \mathbb{R} \to \mathbb{T}^2 \times (\mathbb{R}/\mathbb{Z}) = \mathbb{T}^3.$$

$$p_*(\vec{h}(X^*)) = p_*(\vec{h}(X_3)) - p_*\vec{h}(B') + p_*\vec{h}(B'')$$
$$= p_*\vec{h}(X_3) \quad \text{(again the last two terms add up to 0 as chains).}$$

So

$$p_*\vec{h}(X^*)) = p_*\vec{h}(X_3) = A(X) \quad \text{in } \mathbb{T}^3$$

(we just use all the h_i as $\mathbb{R}/\mathbb{Z}$ valued functions). Further, as $\vec{h}(X^*) = $ boundary of a 3-chain D^* in $\mathbb{T}^2 \times \mathbb{R}$, $p_*\vec{h}(X^*) = A(X)$ bounds $D = p(D^*)$ in $\mathbb{T}^3$ and the harmonic volume

$$\int_D \xi_1 \wedge \xi_2 \wedge \xi_3 = \int_{D^*} \xi_1 \wedge \xi_2 \wedge dx_3$$

where we denote $p^*\xi_3 = dx_3$, $x_3 = $ coordinate on $\mathbb{R}$. By Stokes, the harmonic volume is

$$\int_{\vec{h}(X^*)} x_3\xi_1 \wedge \xi_2 \;\; = \int_{\vec{h}(X_3)} x_3\xi_1 \wedge \xi_2 - \int_{\vec{h}(B')} x_3\xi_1 \wedge \xi_2 + \int_{\vec{h}(B'')} x_3\xi_1 \wedge \xi_2$$

(integration in $\mathbb{T}^2 \times \mathbb{R}$). Since $\vec{h}(B'')$ is just $\vec{h}(B')$ with x_3 replaced by $x_3 + 1$, (and ξ_1, ξ_2 are unchanged by this translation), the last two terms add to

$$\int_{\vec{h}(B'')} \xi_1 \wedge \xi_2 = \int_{B''} \vec{h}^*(\xi_1) \wedge \vec{h}^*(\xi_2).$$

Since B'' is a disk, $\vec{h}^*(\xi_1)$ is exact on it; on the boundary C_3'', $\vec{h}^*(\xi_1) = \alpha_1$ and so α_1 extends to an exact form, say df_1 on B''. Then

$$\int_{B''} \vec{h}^*(\xi_1) \wedge \vec{h}^*(\xi_2) = \int_{C_3''} f_1 \vec{h}^*(\xi_2)$$

$$= \int_{C_3''} f_1 \alpha_2 = \int_{C_3''} (\alpha_1, \alpha_2),$$

the iterated integral. Note that f_1 is only unique up to adding a constant, but $\int_{C_3''} \alpha_2 = \int_{C_3} \alpha_2 = 0$.

Finally, the last two terms add to $\int_{C_3} (\alpha_1, \alpha_2)$. The first term equals $\int_{X_3} h_3 \alpha_1 \wedge \alpha_2$, which we have shown to equal $\int_{C_3} \alpha_{1,2}$. So the harmonic volume equals the iterated integral.

2.4 Use of the Jacobian

Recall that (see for instance [Griffiths and Harris]) the Jacobian manifold $Jac(X)$ can be defined as the compact torus $H_1(X; \mathbb{R})/H_1(X; \mathbb{Z})$ and the Abel-Jacobi map $A : X \to Jac(X)$, taking a base point $x_0 \in X$ to the identity element, is just the iterated integral map II using only the Chen connection modulo commutators of $\mathfrak{g}$, *i.e.*, using the commutative Lie algebra $H_1(X; \mathbb{R})$ and the connection $\sum_{i=1}^{2g} \alpha_i \otimes h_i$ where $\{\alpha_i\}$ is a basis of $\mathcal{H}_{\mathbb{Z}}^1(X)$ and h_i the vector space dual basis of $H_1(X; \mathbb{Z})$.

Let X^- denote the image of $A(X)$ (denoted also by X), under the map $g \to g^{-1}$ on the torus $Jac(X)$. It is immediate that this map $g \to g^{-1}$ takes any translation invariant 1-form on $Jac(X)$ to its negative, and so is the identity on translation invariant two forms, which are the harmonic forms in a translation-invariant metric. Thus both $X (i.e.\ A(X))$ and X^- are cycles representing the same 2-dimensional homology class on $Jac(X)$ and so their difference is a bounding 2-cycle: $X - X^- = \partial E$, E a 3-chain (unique only modulo integral 3-cycles).

The concept of the Jacobian variety $Jac(X)$ can be extended to involve higher dimensional forms and cycles on a Riemannian manifold, for instance $Y = Jac(X)$. Given the 2-cycle $Z = X - X^-$ which is homologous to 0, we choose as before a 3-chain E such that $Z = \partial E$ and consider all harmonic

3-forms β on $Jac(X)$. The choice of E now defines a linear function

$$\mathcal{H}^3(Jac(X)) \to \mathbb{R}, \quad \beta \mapsto \int_E \beta.$$

However E is only well-defined modulo integral 3-cycles, so that associated with Z we obtain a homomorphism

$$\nu(Z): \mathcal{H}^3(Jac(X)) \to \mathbb{R}/\mathbb{Z}.$$

$\nu(Z)$ is said to be a point on this intermediate Jacobian (consisting of all homomorphisms to $\mathbb{R}/\mathbb{Z}$).

We will describe the effect of $\nu(Z)$ on a harmonic(=translation invariant) 3-form β in $H^3(Jac(X); \mathbb{Z})$, which is isomorphic to $\Lambda^3 H^1(X; \mathbb{Z})$, only in the case that β is in the subgroup $\Lambda^3 H^1(X; \mathbb{Z})' = P$, considered above. However P is in a sense the most important part of $H^3(Jac(X); \mathbb{Z})$:

Proposition 2.1   *(see [Harris 1983]) As subgroup of $H^3(Jac(X); \mathbb{Z})$, P is the* primitive *subgroup in the sense of Lefschetz, that is it equals the kernel of the homomorphism*

$$H^3(Jac(X); \mathbb{Z}) \to H^{2g-1}(Jac(X); \mathbb{Z})$$

given by multiplication by ω^{g-2}, where ω is the Kähler form on $Jac(X)$.

Our main result is now

Theorem 2.2   *Let X be a compact connected Riemann surface of genus ≥ 3, $Jac(X)$ its Jacobian, $\nu(X - X^-)$ the point on the intermediate Jacobian of $Jac(X)$ corresponding to primitive harmonic 3-forms. Let $X - X^- = \partial E$ on $Jac(X)$ and let $\alpha_1, \alpha_2, \alpha_3 \in \mathcal{H}^1_{\mathbb{Z}}(X)$ be harmonic forms Poincaré dual to disjoint simple closed curves C_1, C_2, C_3 on X. Regarding the α_i also as 1-forms on $Jac(X)$, we have*

$$\nu(X - X^-)(\alpha_1 \wedge \alpha_2 \wedge \alpha_3) = \int_E \alpha_1 \wedge \alpha_2 \wedge \alpha_3 \qquad \mathrm{mod}\ \mathbb{Z}$$

$$= 2 \int_{C_3} (\alpha_1, \alpha_2) + \alpha_{12} \qquad \mathrm{mod}\ \mathbb{Z}$$

(α_{12}=unique coexact 1-form on X such that $\alpha_1 \wedge \alpha_2 + d\alpha_{12} = 0$).

Proof: We use $Jac(X) \xrightarrow{\pi} \mathbb{T}^3$, the natural projection (dual to the inclusion $\mathbb{Z}\alpha_1 + \mathbb{Z}\alpha_2 + \mathbb{Z}\alpha_3 \subset \mathcal{H}^1_{\mathbb{Z}}(Jac(X))$. Note that π commutes with the (-1) map, $g \to g^{-1}$. For the 3-chain E in $Jac(X)$, $\pi(E)$ satisfies

$$\partial\pi(E) = \pi(X) - \pi(X^-) = \partial(D - D^-)$$

where D^- =image of D under (-1) map of $\mathbb{T}^3$. See Definition 2.1 for D. Thus

$$\int_{\pi(E)} \xi_1 \wedge \xi_2 \wedge \xi_3 = \int_D \xi_1 \wedge \xi_2 \wedge \xi_3 - \int_{D^-} \xi_1 \wedge \xi_2 \wedge \xi_3 \qquad \text{mod } \mathbb{Z}$$

and

$$\int_{D^-} \xi_1 \wedge \xi_2 \wedge \xi_3 = -\int_D \xi_1 \wedge \xi_2 \wedge \xi_3$$

since the (-1) map sends $\xi_1 \wedge \xi_2 \wedge \xi_3$ to its negative. This concludes the proof. $\square$

In summary we have assigned to a compact complex Riemann surface X a point $\nu(X)$ on the intermediate Jacobian of $Jac(X)$ given by integration of primitive harmonic 3-forms on $J(X) = Jac(X)$ over any 3-chain in $J(X)$ whose boundary is $X - X^-$, and have shown that $\nu(X) = 2 \times$ harmonic volume $= 2 \times$ iterated integral.

However we have not yet shown that these quantities can be non-zero, (for $g \geq 3$) and so we will prove a variational formula for them (their derivatives with respect to moduli), which will in particular prove they do not vanish identically.

2.5   Variational formula for harmonic volume

We will work over Torelli space , defined as the set of equivalence classes of pairs (compact Riemann surface X, symplectic basis for $H_1(X; \mathbb{Z})$ relative to intersection number) where equivalence is given by complex analytic homeomorphism taking the given symplectic bases into one another. We will define a neighborhood of a pair by a type of deformation of the complex structure called a "Schiffer variation". Torelli space is a quotient space of Teichmuller space by the action of a discrete group of complex analytic transformations acting freely, and is a complex manifold of complex dimension $3g - 3$.

A Schiffer variation of complex structure(described in [Schiffer and Spencer], Chapter 8, §1) is described as follows: fix a point $t \in X$ and a local coordinate z in a neighborhood of t with $z(t) = 0$. Let z^* be a local coordinate in a region of X which overlaps the domain of z in an annulus but excludes the disk $|z| \leq \rho$.

X is obtained by identifying the domains of z and z^* on their overlap by the identity map $z^* = z$, while a new surface X^* is obtained by the new identification

$$z^* = z + (s/z)$$

where $s = e^{2i\phi}\rho^2$ is any sufficiently small complex number. X corresponds to $s = 0$.

A vector field in the neighborhood of the point t (*i.e*, of $z = 0$) is given by $(1/z)d/dz$(which is also expressible as $\partial/\partial s$ at $s = 0$). This (meromorphic) vector field can be regarded as an element of $H^1(X, K^{-1})$, K=canonical bundle (*i.e.*, cotangent bundle) of X. The vector space $H^1(X, K^{-1})$ is dual to the vector space $H^0(X, K^2)$ of holomorphic quadratic differentials $q(z)(dz)^2$ on X, the pairing being given by the residue at t:

$$(q(z)(dz)^2, \frac{1}{z}d/dz) = Res_t q(z)(dz)^2(\frac{1}{z}d/dz)$$

$$= Res_t q(z)\frac{dz}{z} = q(t).$$

The variation of complex structure just described is not the most general one, so we generalize it by first choosing on X $3g - 3$ *general* distinct points t_i, $i = 1, \cdots, 3g - 3$; this means that we choose the t_i so that any quadratic differential vanishing at all of them must be zero. We will prove a little later that such (distinct) points $t_1, \cdots, t_{3g-3}$ exist. Then we choose sufficiently small disks around the t_i so that they are disjoint, and use gluing maps

$$z_i^* = z_i + (s_i/z_i).$$

This gives us a $3g - 3$ dimensional family of deformations parametrized by small $s = (s_1, \cdots, s_{3g-3})$. The Kodaira-Spencer theory (Kodaira-Morrow, "Complex Manifolds", Chapter 2, Section 3) tells us that we obtain in this way all deformations sufficiently near X, *i.e.*, a neighborhood in Torelli (or Teichmuller) space, whose tangent space at the point given by X is $H^1(X; K^{-1}) =$ dual space to $H^0(X; K^2)$.

We can now regard the Abel-Jacobi point $\nu(X_s)$ as a function of s with values in the intermediate Jacobian given by primitive 3-forms $P(X_s)$. If we give these intermediate Jacobians the right complex structure (following Griffiths), ν becomes a holomorpic section of a complex analytic family of complex tori.

To compute the variation of harmonic volume with s, we use formulas of Schiffer and Spencer to see how a set of harmonic forms $\alpha_1, \alpha_2, \alpha_3$ vary with s (see [Harris 1983a], Sect. 5). Let the map $X \to (\mathbb{R}/\mathbb{Z})^3$ used to define harmonic volume be denoted

$$\vec{h} = (h_1, h_2, h_3)$$

(so $\alpha_i = dh_i$). Let X^* be the "nearby" deformed surface with corresponding $\vec{h}^*$ which is close to $\vec{h}$. We need to calculate to first order in s the volume between the surfaces $\vec{h}^*(X)$ and $\vec{h}(X)$ in $\mathbb{T}^3 = (\mathbb{R}/\mathbb{Z})^3$.

This volume is a sum of volumes of small parallelepipdes with bases $d\vec{h} \times d\vec{h}$ in $\vec{h}(X)$ and (slanted) heights $\vec{h}^* - \vec{h}$. Here $d\vec{h} \times d\vec{h}$ denotes the cross-product of 3-dimensional vectors(*i.e.* the Lie bracket in the Lie algebra of $SO(3)$) combined with the wedge product of 1-forms. The volume element is thus the dot product

$$\frac{1}{2}(\vec{h}^* - \vec{h}) \cdot (d\vec{h} \times d\vec{h}).$$

The formulas of Schiffer and Spencer lead us to the following calculation: as before, $dh_i = \alpha_i$ and there is a unique coexact form α_{ij} so that

$$\alpha_i \wedge \alpha_j + d\alpha_{ij} = 0 \quad \text{on } X.$$

Let

$$\alpha_{ij} + i \star \alpha_{ij} = a_{ij}(z, \bar{z})dz$$

be the (non-holomorphic) 1-form of type (1, 0) with real part α_{ij}, and similarly let

$$\alpha_k + i \star \alpha_k = b_k(z)dz$$

be the holomorphic 1-form(*i.e.*, of type (1, 0)).

Consider the quadratic differential obtained by multiplying these (symmetric product) for distinct i, j, k:

$$(\alpha_{ij} + i \star \alpha_{ij})(\alpha_k + i \star \alpha_k) = a_{ij}b_k(dz)^2.$$

This is a non-holomorphic quadratic differential $q_{ijk}(z)(dz)^2 = Q_{ijk}$. We write $q_{ijk} = Q_{ijk}/(dz)^2$.

However, minus the sum of these over the 3 *cyclic permutations* of $(i, j, k) = (1, 2, 3)$, *i.e.*, $Q(\alpha_1, \alpha_2, \alpha_3)$ defined as

$$- \sum_{(1,2,3)} (\alpha_{12} + i \star \alpha_{12})(\alpha_3 + i \star \alpha_3)$$

is holomorphic, and Q defines an $\mathbb{R}$-linear map on $\Lambda^3 \mathcal{H}^1_{\mathbb{Z}}(X)' \otimes_{\mathbb{Z}} \mathbb{R} = P \otimes_{\mathbb{Z}} \mathbb{R}$:

$$Q : P \otimes_{\mathbb{Z}} \mathbb{R} \to H^0(X; K^2)$$

(where $P \subset \mathcal{H}^3_{\mathbb{Z}}(J)$ as before).

This $\mathbb{R}$-linear map is $\mathbb{C}$-linear if we choose the Griffiths complex structures on $P \otimes_{\mathbb{Z}} \mathbb{R}$ (see [Harris, 1983a], Sect.4) which is defined as follows on the larger space $\Lambda^3 \mathcal{H}^1(J; \mathbb{R}) = \Lambda^3 \mathcal{H}^1(X; \mathbb{R})$: first, define a complex structure j_1 on $\mathcal{H}^1(X)$ as $j_1 = -\star$, so $j_1 \otimes 1$ on holomorphic $(1, 0)$ forms is multiplication by i. Then on $\Lambda^3 \mathcal{H}^1(X; \mathbb{R})$ let j_3 be

$$j_3(A \wedge B \wedge C) = \frac{1}{2}[j_1(A) \wedge j_1(B) \wedge j_1(C) + j_1(A) \wedge B \wedge C \\ + A \wedge j_1(B) \wedge C + A \wedge B \wedge j_1(C)].$$

On $(H^{3,0} \oplus H^{2,1})(J)$ this is multiplication by i, and on $H^{1,2} \oplus H^{0,3}$ it is multiplication by $-i$. We then have the following theorem (*loc. cit.* Theorem 5.8)

Theorem 2.3   *Let $\lambda = \alpha_1 \wedge \alpha_2 \wedge \alpha_3 \in P$ and consider the harmonic volume*

$$I(\lambda; s) = I(\alpha_1, \alpha_2, \alpha_3; s)$$

for $s = (s_1, \cdots, s_{3g-3})$ near $s = 0$. Then

$$I(\lambda; s) - I(\lambda; 0) = Imaginary\ part\ of\ [2\pi \sum_{j=1}^{3g-3} s_j \frac{Q(\lambda)}{(dz_j)^2}(t_j)] + o(s)$$

(Here t_j are the $3g-3$ general points on X, with z_j local coordinates centered at t_j and $\frac{Q(\lambda)}{(dz_j)^2}$ the function obtained by dividing the quadratic differentials by $(dz_j)^2$ into which we substitute t_j). $I(\lambda, s)$ is a holomorphic function of s for each λ.

We can explicitly calculate the differential

$$\delta I : P \otimes \mathbb{R} = \Lambda^3 \mathcal{H}^1_{\mathbb{R}}(X)' \to H^0(X; K^2)$$

of I at $s = 0$ on the locus in Torelli space of hyperelliptic Riemann surfaces

$$w^2 = (z - e_1) \cdots (z - e_n), \quad n = 2g + 2, g \geq 3$$

if the branch points $e_1, \cdots, e_n$ (distinct complex numbers) satisfy

$$\prod_{i=1}^{n} |z - e_i|$$

is a symmetric function of $\mathrm{Arg}z$, and prove that in this case

$$\delta I : \Lambda^3 \mathcal{H}_{\mathbb{R}}^1(X)' \to H^0(X; K^2)$$

has as image the whole -1 eigenspace of the hyperelliptic involution, *i.e.*, this image is as large as possible [Harris 1983a].

   To summarize, if X is hyperelliptic then $\nu = 2I$ vanishes on $X - X^-$ (since $X^- = X$) and so I takes on values 0 or $\frac{1}{2}$ mod $\mathbb{Z}$, and is therefore constant as $\mathbb{R}/\mathbb{Z}$ valued function on the (connected) hyperelliptic locus, whose complex dimension is $2g - 1$.

   The space of tangent vectors to Torelli space at (X) which are "normal" to the hyperelliptic locus is $g-2$ dimensional and dual to the (-1) eigenspace of the hyperelliptic involution of X acting on the quadratic differentials of X. Regarding $I = I_s$ as function of the moduli parameters s (and constant on the hyperelliptic locus) and regarding its differential δI as linear mapping on the (-1) eigenspace, δI is injective on this normal space, if X varies in a non-empty open subset of the hyperelliptic locus.

   More precisely, Torelli space is a complex manifold and over it we have the family of complex tori $Hom(P, \mathbb{R}/\mathbb{Z})$ which is a complex vector bundle modulo a locally constant family of lattices (certain integral coefficient homology groups). Iterated integrals define a holomorphic section of the vector bundle whose differential is injective in the normal direction to an open subset of a submanifold (along which the section vanishes). In particular the section of the bundle of complex tori is not identically zero.

   We still have to prove the existence of divisors $D = t_1 + \cdots + t_{3g-3}$, with distinct t_i, such that any holomorphic quadratic differential vanishing at every t_i must be 0, in other words

$$H^0(K^2 D^{-1}) = (0).$$

   The degrees of these divisors are as follows: $\deg K = 2g - 2$, $\deg K^2 = 4g - 4$, $\deg K^2 D^{-1} = g - 1$. By Riemann-Roch, the dimensions satisfy

$$h^0(K^2 D^{-1}) - h^0(DK^{-1}) = g - 1 - (g - 1) = 0.$$

Thus we must find D so that $h^0(DK^{-1}) = 0$: this means that in the linear equivalence class of DK^{-1} (which has degree $g - 1$) there is no effective divisor D'. Equivalently, D is not linearly equivalent to KD' for any effective divisor D' (of degree $g - 1$). Consider now the (Abel-Jacobi) map u from divisors of degree $3g - 3$ to the Picard variety Pic^{3g-3} of isomorphism classes of line bundles of degree $3g - 3$ (isomorphic to $Jac(X)$):

$$u : Sym^{3g-3}(X) \to Pic^{3g-3}.$$

All fibers of u are projective spaces P^{2g-3}. Within Pic^{3g-3} we have the subvariety given by all line bundles KD' ($\deg D' = g - 1$), which is $u(K \ Sym^{g-1}(X))$ and so has dimension $\leq g - 1$. Thus we just need to choose D in the open subset of $Sym^{3g-3}(X)$ such that $u(D) \in Pic^{3g-3} - u(K \ Sym^{g-1})$ and none of the $3g - 3$ points are equal.

So we have proved the existence of the "general" divisors D, and the nontriviality of the section I of the family of primitive intermediate Jacobians, taking a point corresponding to X to the Abel-Jacobi image $\nu(X)$. However this result does not directly exhibit a Riemann surface X where this Abel-Jacobi image $\nu(X)$ is non-zero. A little further on in this chapter we will exhibit such an X, namely a well-known algebraic curve (of genus 3)

$$x^4 + y^4 = 1$$

defined over $\mathbb{Z}$: to do this we will just calculate some iterated integrals.

In the next section we recall the connection (due to Hodge and Weil) of the Abel-Jacobi map and equivalence relations on algebraic cycles, then give the application of our methods to this problem.

2.6   Algebraic equivalence and homological equivalence of algebraic cycles

We want to look at the problem of whether a complex algebraic 1-cycle Z *i.e.*, of complex dimension 1 on a smooth projective complex algebraic variety V is algebraically equivalent to 0 if it is known to be homologous to 0. We will only give a rather vague, intuitive idea of what algebraic equivalence to 0 means - for more details, see Fulton's "Intersection Theory". We will use the following definition: suppose the complex 1-cycle Z (considered as topological 2-cycle) is the boundary of a topological 3-chain W (*i.e.*, W has topological dimension 3). We then say that Z is homologous to 0. Suppose further there is an algebraic (or complex analytic) subset S of V of complex dimension 2 such that the topological 3-chain W lies on S: we then say that Z is algebraically equivalent to 0 (in V). A more intuitively understandable situation is the following: suppose Z_0, Z_1 are two irreducible one-dimensional complex subvarieties of V which are two members of a *family* of subvarieties Z_t, $t \in T$, all Z_t being 1-dimensional (over $\mathbb{C}$) and T being a *connected* algebraic curve over $\mathbb{C}$. In this latter case the union, in V, of the Z_t will be the subset S of complex dimension 2. (If further the parameter space T is a rational curve then $Z_0, Z_!$ are said to be rationally equivalent.) Here we let $Z = Z_1 - Z_0$.

Hodge and Weil (see Weil, Collected Papers, Paper 1952e and comments on it) proposed the following criterion for algebraic equaivalence to 0 of the complex 1-cycle Z on V - the criterion is clearly necessary but it is not obviously sufficient: if $Z = \partial W$ on S then for every complex $(3, 0)$ form ω on V, the restriction of ω to S is clearly 0 (since S has complex dimension 2) and so $\int_W \omega = 0$. However since the chain W is not unique but is only unique modulo topological 3-cycles, the criterion is that if Z is algebraically equivalent to 0 then the linear function it defines on complex harmonic forms of type (3,0) is 0 modulo the linear functions of integration over all topological 3-cycles.

Denoting real harmonic primitive 3-forms by P^3, we form $P^3 \otimes_{\mathbb{R}} \mathbb{C}$ and use the Hodge decomposition

$$P^3 \otimes \mathbb{C} = P^{3,0} \oplus P^{2,1} \oplus P^{1,2} \oplus P^{0,3}.$$

In fact $P^{3,0} = \mathcal{H}^{3,0}$, $P^{0,3} = \mathcal{H}^{0,3}$ because $\omega \in \mathcal{H}^{1,1}$.

For example, if $V = Jac(X)$, X a genus 3 curve, then $\mathcal{H}^{1,0}$ and $\mathcal{H}^{0,1}$ each have dimension 3 over $\mathbb{C}$, $\mathcal{H}^{3,0}$ and $\mathcal{H}^{0,3}$ are each 1-dimensional, and $\mathcal{H}^{2,1}$ and $\mathcal{H}^{1,2}$ are each 9-dimensional. P^3 has dimension 14, $P^{3,0} = \mathcal{H}^{3,0}$ and $P^{0,3} = \mathcal{H}^{0,3}$ are 1-dimensional. The primitive intermediate Jacobian of X we use is ($*$ denoting complex dual)

$$\frac{(P^{3,0} \oplus P^{2,1})^*}{\text{Image } H_3(Jac(X); \mathbb{Z})}$$

and we have an Abel-Jacobi type homomorphism

$$\frac{\text{Complex 1-cycles on } Jac(X), \text{ homologous to } 0}{\text{Those algebraically equivalent to } 0} \to \frac{(P^{3,0})^*}{\text{Image } H_3(Jac(X); \mathbb{Z})}.$$

In general, the image of $H_3(Jac(X); \mathbb{Z})$ in the one dimensional complex vector space $(P^{3,0})^*$ (for X of genus 3) need not be discrete. However we would like to find examples of X such that this image is discrete: this will be the case if for a basis γ_k of $H_1(X; \mathbb{Z}$ and a $\mathbb{C}$-basis θ_j of $\Omega^{1,0}(X)$ (holomorphic 1-forms), the period matrix, with entries $\int_{\gamma_k} \theta_j$, has entries in a discrete subring of $\mathbb{C}$. Then on $J(X)$ the (3,0) form $\theta_1 \wedge \theta_2 \wedge \theta_3$ has integral over each 3-cycle a 3×3 minor (determinant) of the 3×6 period matrix and this is again in the discrete subring of $\mathbb{C}$. We can then try to calculate the image of the 3-cycle $X - X^-$ in $(P^{3,0})^* /$ Image $H_3(Jac(X); \mathbb{Z})$ and see if it is non-zero.

This program can be carried ot for certain X with large automorphism group, for instance the Fermat quartic

$$x^4 + y^4 = 1$$

using formulas for the period matrix entries in a paper of B. Gross and D. Rohrlich (the entries are given by the Gamma function). The details of the calculation will be given in the next section.

Finally we find that

$$\nu(\theta_1 \wedge \theta_2 \wedge \theta_3) = 2I(\theta_1, \theta_2, \theta_3)$$

is non -zero in the group

$$\frac{(P^{3,0})^*}{\text{Image } H_3} = \frac{\mathbb{C}}{\mathbb{Z}(i)}$$

and so $X - X^-$ is not algebraically equivalent to 0. However, further work by S. Bloch, [Bloch], using l-adic cohomology, was needed to prove that $X - X^-$ has no (non-zero) integer multiple which is algebraically equivalent to 0.

It had previously been shown by P.Griffiths using a variational method, that in certain families of cycles homologous to 0, all but some subset of measure 0 of this family of cycles were not algebraically equivalent to 0. However this was basically an existence result, and could not exhibit particular cycles, much less ones defined over $\mathbb{Z}$ or a number field. It was thus important to exhibit examples such as the Fermat quartic on which arithmetic conjectures of Bloch about the "Griffiths group" could be tested.

Iterated integrals on this particular curve are further studied in [Harris, 1990] where the curve is instead described as the modular curve $X_0(64)$ and its holomorphic 1-forms are certain theta functions. Iterated integrals of a class of theta functions, those having harmonic polynomial coefficients, are there shown to be a more general class of functions: theta functions with rational function coefficients. These latter are not modular forms but their Mellin transforms satisfy a certain type of functional equation.

2.7 Calculations for the degree 4 Fermat Curve

This section will reproduce an unpublished paper giving the calculation announced in [Harris, 1983b]. The main result of this calculation is that $\nu(\theta_1 \wedge \theta_2 \wedge \theta_3)$ in the group $\mathbb{C}/\mathbb{Z}(i)$ is (up to multiplication by $-i$), the real number (mod $\mathbb{Z}$)

$$2 \int_0^1 ((1 - x^4)^{-1/2} dx, (1 - x^4)^{-3/4} dx) / \int_0^1 (1 - x^4)^{-1/2} dx \int_0^1 (1 - x^4)^{-3/4} dx$$

where the numerator is an iterated integral. Since the numerator is an integral over a triangle $0 \leq x \leq y \leq 1$ and the denominator is over the square $0 \leq x \leq 1$, $0 \leq y \leq 1$, it is easy to estimate that the real number above is not in $\mathbb{Z}$ Further work by S. Bloch, C. Schoen, and D. Zehinsky has been done to study this and related examples by various techniques of arithnetic geometry (see [Bloch] for an initial paper).

Homological versus algebraic equivalence in a Jacobian

We will consider the problem of deciding whether two homologous algebraic p-cycles C, C' in a compact non-singular algebraic variety V over $\mathbb{C}$ are algebraically equivalent (*i.e.*, roughly speaking, the cycle $C - C'$ can be "continuously" (algebraically) deformed into the cycle 0; see [3] for a precise definition). A direct method of proving algebraic non-equivalence (for $p > 0$) is to consider a singular chain D of (topological) dimension $2p + 1$ whose boundary is $C - C'$ and integrate holomorphic $(2p + 1)$-forms over D thus obtaining an element of the quotient group of the dual of the $(2p + 1)$-forms by the subgroup of periods (integrals over $(2p + 1)$-singular cycles). If this element of the quotient group is non-zero then C, C' are not algebraically equivalent (compare with Hodge's letter of 1951 given in [6]). We carry this out on the example of the Fermat curve $F : x^4 + y^4 = 1$ for C, its Jacobian J for V, and for C' the image of C under the map of J given by the group-structure inverse. The method for calculating integrals over the 3-chains D is that given in our paper [5] and amounts to calculating iterated integrals of holomorphic 1-forms on the Riemann surface C. The special feature of the Fermat curve of degree 4 is that its normalized period matrix has entries in a quadratic imaginary field and so the periods of holomorphic 3-forms on J generate a discrete subgroup of $\mathbb{C}$: consequently, the integrals in question need only be calculated to sufficient accuracy to see that they are not in this discrete group. Thus the same method can, in principle, be applied to any curve with normalized period matrix in a quadratic imaginary field.

The above direct method has not been carried out before now, perhaps because of lack of a formula for the integrals involved. Instead, Griffiths([2]) developed another method, by which he gave the first examples of homologous non-algebraically equivalent cycles, namely consideration of families V_t of varieties depending on parameter t, and differentiation with respect to these parameters. However, this method only allows one to prove that a *generic* member V_t of the family carries such cycles (*generic* usually meaning that t lies in the complement of a countable union of proper subvarieties). All subsequent examples of algebraic non-equivalence have followed the Griffiths method. In particular, Ceresa [1] recently proved algebraic non-euqivalence for a non-singular *generic* curve C and $C' = $ inverse (C) in $J(C)$ by this differential method; our paper [5] contains another way of doing this. The problem for curves C over Q was mentioned to us by Spencer Bloch; I wish to thank him, Ron Donagi and Gerry Washnitzer for conversations.

1. Let C be a non-singular curve of genus ≥ 3, and let $H^{p,q}$ denote the $\mathbb{C}$-space of harmonic (p, q) forms on $J(C)$. Let P_+ denote the subspace of *primitive* forms in $H^{3,0} + H^{2,1}$, so that $H^{3,0} \subset P_+$, P_+^* its complex dual, and L the lattice in P_+^* of all linear functions obtained by integrating over integral 3-cycles in J. P_+^*/L is a complex torus and the cycle $C - C'$ determines a point ν on this torus, namely $\nu(\phi) = \int_D \phi$ (mod "periods"), for $\phi \in P$, where $\partial D = C - C'$. The embedding of C in J, and hence of C', depends on the choice of a base point $p_0 \in C$, but ν is independent of base point because we deal only with primitive 3-forms.

We recall from [5] how to calculate ν in terms of iterated integrals in C of real harmonic 1-forms α, β: if $\alpha \wedge \beta$ is an exact 2-form on C, $\alpha \wedge \beta = d\eta$ where η is uniquely determined by being required to be orthogonal to all closed 1-forms on C, and if l is a path, parametrized by $t \in [0, 1]$, then the iterated integral is

$$I(\alpha, \beta; l) = \int_{x=0}^{1} [\int_{t=0}^{x} \alpha(t)] \beta(x) - \eta(x). \qquad (2.1)$$

$I(\alpha, \beta; l)$ depends only on the homotopy class l (with fixed endpoints). If $l \cdot l'$ denotes a product path, l^{-1} the path inverse to l, and c a closed path then

$$I(\alpha, \beta; l \cdot l') = I(\alpha, \beta; l) + I(\alpha, \beta; l') + \int_l \alpha \cdot \int_{l'} \beta \qquad (2.2)$$

$$I(\alpha, \beta; l \cdot l^{-1}) = 0 \qquad (2.3)$$

$$I(\alpha, \beta; l \cdot c \cdot l^{-1}) = I(\alpha, \beta; c) + \int_l \alpha \cdot \int_c \beta - \int_l \beta \cdot \int_c \alpha \qquad (2.4)$$

$$I(\alpha, \beta; l) + I(\beta, \alpha; l) = \int_l \alpha \int_l \beta \qquad (2.5)$$

(α, β) may be replaced by any element of $(\mathcal{H} \otimes_{\mathbb{R}} \mathcal{H})'$ where $\mathcal{H}$ denotes real harmonic 1-forms and $(\mathcal{H} \otimes \mathcal{H})'$ is the subspace of $\mathcal{H} \otimes \mathcal{H}$ annihilated by the real linear function $\int_c \alpha \wedge \beta$. Let $\mathcal{H}_{\mathbb{Z}}$ denote real harmonic 1-forms with periods in $\mathbb{Z}$ and $(\mathcal{H} \otimes_{\mathbb{Z}} \mathcal{H})'$ the corresponding subspace of $\mathcal{H}_{\mathbb{Z}} \otimes \mathcal{H}_{\mathbb{Z}}$.

Fix a base point $p_o \in C$ and let $\pi_1(C, p_0)^{ab}$ be the abelianized fundamental group, isomorphic to $\mathcal{H}_{\mathbb{Z}}$ under Poincaré duality. Then (2.1), (2.2) allows us to define a homomorphism

$$\bar{I} : (\mathcal{H}_{\mathbb{Z}} \otimes \mathcal{H}_{\mathbb{Z}})' \otimes \pi_1(C, p_0)^{ab} \to \mathbb{R}/\mathbb{Z} \qquad (2.6)$$

equivalently: $(\mathcal{H}_{\mathbb{Z}} \otimes \mathcal{H}_{\mathbb{Z}})' \otimes \mathcal{H}_{\mathbb{Z}} \to \mathbb{R}/\mathbb{Z}$, $(\bar{I}(\alpha, \beta; C) = \int_C [(\int \alpha)\beta - \eta]$ mod $\mathbb{Z}$).

Let $\mathbb{Z}(i)$ denote the Gaussian integers, $\mathcal{H}_{\mathbb{Z}(i)} = \mathcal{H} \otimes \mathbb{Z}(i)$ the complex-valued harmonic 1-forms with periods in $\mathbb{Z}(i)$, and $H^{1,0}_{\mathbb{Z}(i)}$ the $\mathbb{Z}(i)$ submodule of holomorphic 1-forms with periods in $\mathbb{Z}(i)$. On tensoring (2.6) with $\mathbb{Z}(i)$ we get

$$\bar{I} : (\mathcal{H}_{\mathbb{Z}(i)} \otimes_{\mathbb{Z}(i)} \mathcal{H}_{\mathbb{Z}(i)})' \otimes_{\mathbb{Z}(i)} (\pi_1(C, p_0)^{ab} \otimes \mathbb{Z}(i)) \to \mathbb{C}/\mathbb{Z}(i) \qquad (2.7)$$

$$or : (\mathcal{H}_{\mathbb{Z}(i)} \otimes_{\mathbb{Z}(i)} \mathcal{H}_{\mathbb{Z}(i)})' \otimes_{\mathbb{Z}(i)} \mathcal{H}_{\mathbb{Z}(i)} \to \mathbb{C}/\mathbb{Z}(i). \qquad (2.8)$$

In [5], Sec. 2, we prove the following: in the commutative diagram with exact rows

$$
\begin{array}{ccccccccc}
0 & \longrightarrow & (\mathcal{H}_{\mathbb{Z}} \otimes \mathcal{H}_{\mathbb{Z}} \otimes \mathcal{H}_{\mathbb{Z}})' & \longrightarrow & \mathcal{H}_{\mathbb{Z}} \otimes \mathcal{H}_{\mathbb{Z}} \otimes \mathcal{H}_{\mathbb{Z}} & \xrightarrow{\tilde{P}} & \mathcal{H}_{\mathbb{Z}} \oplus \mathcal{H}_{\mathbb{Z}} \oplus \mathcal{H}_{\mathbb{Z}} & \longrightarrow & 0 \\
& & \downarrow j_1 & & \downarrow j_2 & & \downarrow j_3 & & \\
0 & \longrightarrow & P_{\mathbb{Z}} & \longrightarrow & \Lambda^3(\mathcal{H}_{\mathbb{Z}}) & \xrightarrow{\bar{P}} & \mathcal{H}_{\mathbb{Z}} & \longrightarrow & 0
\end{array}
$$

where, denoting $\int_C x \wedge y$ by $x \cdot y$ for $x, y \in \mathcal{H}_{\mathbb{Z}}$,

$$\tilde{P}(x \otimes y \otimes z) = (x \cdot y)z \oplus (y \cdot z)x \oplus (z \cdot x)y$$

$$\bar{P}(x \wedge y \wedge z) = (x \cdot y)z + (y \cdot z)x + (z \cdot x)y$$

j_2 is the natural homomorphism and induces j_1) we have: j_1 is surjective, $(\mathcal{H}_{\mathbb{Z}} \otimes \mathcal{H}_{\mathbb{Z}} \otimes \mathcal{H}_{\mathbb{Z}})'$ is a subgroup of the domain of $\bar{I}$, and $2\bar{I}$ restricted to

$(\mathcal{H}_{\mathbb{Z}} \otimes \mathcal{H}_{\mathbb{Z}} \otimes \mathcal{H}_{\mathbb{Z}})'$ factors through j_1: there is a unique homomorphism $\bar{\nu} : P \to \mathbb{R}/\mathbb{Z}$ with $\bar{\nu} \circ j_1 = 2\bar{I}$ and a commutative diagram

$$(\mathcal{H}_{\mathbb{Z}} \otimes \mathcal{H}_{\mathbb{Z}} \otimes \mathcal{H}_{\mathbb{Z}})' \subset (\mathcal{H}_{\mathbb{Z}} \otimes \mathcal{H}_{\mathbb{Z}})' \otimes \mathcal{H}_{\mathbb{Z}} \xrightarrow{2\bar{I}} \mathbb{R}/\mathbb{Z}$$

$$\searrow \qquad \nearrow \bar{\nu} \qquad\qquad (2.9)$$

$$P$$

and on extending scalars to $\mathbb{Z}(i)$

$$\bar{\nu} : P_{\mathbb{Z}(i)} \to \mathbb{C}/\mathbb{Z}(i) \qquad\qquad (2.10)$$

(2.9), (2.10) can be rewritten as

$$\bar{\nu} \in \mathrm{Hom}_{\mathbb{R}}(P_{\mathbb{Z}} \otimes \mathbb{R}, \mathbb{R})/\mathrm{Hom}(P_{\mathbb{Z}}, \mathbb{Z}) \subset \mathrm{Hom}_{\mathbb{C}}(P, \mathbb{C})/\mathrm{Hom}_{\mathbb{Z}(i)}(P_{\mathbb{Z}(i)}, \mathbb{Z}(i)).$$

$\mathrm{Hom}(P_{\mathbb{Z}}, \mathbb{Z})$ is the restriction to $P_{\mathbb{Z}}$ of $H_3(J; \mathbb{Z}) = \mathrm{Hom}(H^3(J; \mathbb{Z}), \mathbb{Z})$ (since $P_{\mathbb{Z}}$ is a direct summand of $H^3(J; \mathbb{Z})$), and similarly, $\mathrm{Hom}_{\mathbb{Z}(i)}(P_{\mathbb{Z}(i)}, \mathbb{Z}(i))$ is the restri tion of $H_3(J; \mathbb{Z}(i))$. It is shown in [6] that the ν of the beginning of this section is the same as the $\bar{\nu}$ of (2.9), (2.10). From now on, we will write only ν or I for these homomorphisms.

We will only be interested in the restriction of ν of (2.10) to certain holomorphic 3-forms, namely $\Lambda^3_{\mathbb{Z}(i)}(H^{1,0}_{\mathbb{Z}(i)}) \subset \Lambda^3_{\mathbb{Z}(i)}(\mathcal{H}_{\mathbb{Z}(i)})'$. *We assume from now on that* $H^{1,0} = H^{1,0}_{\mathbb{Z}(i)} \otimes_{\mathbb{Z}(i)} \mathbb{C}$, and let $\theta_1, \cdots, \theta_g$ be a $\mathbb{Z}(i)$ basis for $H^{1,0}_{\mathbb{Z}(i)}$. Then $\nu(\theta_i \wedge \theta_j \wedge \theta_k) = 2I(\theta_i \otimes \theta_j \otimes \theta_k)$ is calculated as follows: by Poincaré duality, the dual of θ_k is $\sum a_r K_r$ where $K_1, \cdots, K_{2g}$ is a $\mathbb{Z}$-basis of $H_1(C; \mathbb{Z})$ and $a_r \in \mathbb{Z}(i)$. Let K_r be the homology class of $c_r \in \pi_1(C; p_0)$, then $I(\theta_i \otimes \theta_j \otimes \theta_k) = \sum a_r \int_{C_r} (\theta_i, \theta_j)$ ($\mod \mathbb{Z}(i)$). ν is determined only modulo $\mathrm{Hom}_{\mathbb{Z}(i)}(\Lambda^3(H^{1,0}_{\mathbb{Z}(i)}), \mathbb{Z}(i))$, which is generated by the homomorphisms

$$\int_{K_{P_1} \wedge K_{P_2} \wedge K_{P_3}} \theta_{i_1} \wedge \theta_{i_2} \wedge \theta_{i_3} = \det[\int_{K_{P_s}} \theta_{i_r}] \in \mathbb{Z}(i).$$

For simplicity, let $g = 3$, then $\tilde{I}$ is given by one complex number (a linear combination of iterated integrals) modulo $\mathbb{Z}(i)$, and *approximate* evaluation of the integrals will suffice. It is clear that $\mathbb{Z}(i)$ could be replaced by an order R in an imaginary quadratic field, but it is essential that R be discrete in $\mathbb{C}$ if approximate evalution of integrals is used.

2. Let $C = F(4) : x^4 + y^4 = 1$. [4] will be our main reference here.

We consider this Riemann surface as 4-sheeted covering of the x-plane brnached over $x = i^j$, $j = 0, 1, 2, 3$, $(i = e^{i\pi/2})$. Make cuts from i^j radially out to ∞, thus defining four sheets which will be labelled by the index $S = 0, 1, 2, 3$ such that at $X = 0$, $y = i^S$ on sheet S.

Let a be the path going from 1 to i on sheet 1 and returning from i to 1 on sheet 0:

$$a = (1i)_1 \cdot (i1)_0.$$

This is the path used in [4], and its homology class together with its transforms under certain automorphisms give a basis of $H_1(F; \mathbb{Z})$. The automorphisms are A, B:

$$A(x, y) = (ix, y) \quad B(x, y) = (x, iy).$$

A homology basis is

$$K_1 = a, \ K_2 = Aa, \ K_3 = B^2 a, \ K_4 = A^{-1} Ba,$$
$$K_5 = A^2 B^2 a, \ K_6 = A^2 B^{-1} a. \tag{2.11}$$

The intersection numbers are

$$K_{2j-1} \circ K_{2j} = 1, \ K_r \circ K_s = 0 \text{ if } (r, s) \neq (2j - 1, 2j) \text{ or } (2j, 2j - 1).$$

Let $\eta_{r,s,t} = x^{r-1} y^{s-4} dx$. Then $\eta_1 = \eta_{1,1,2}$, $\eta_2 = \eta_{1,2,1}$, $\eta_3 = \eta_{2,1,1}$ are holomorphic and a basis for $H^{1,0}$. Further $A^j B^k \eta_{r,s,t} = i^{jr+ks} \eta_{r,s,t}$. Let

$$\eta_1^* = \eta_{1,1,2}/B(\frac{1}{4}, \frac{1}{4})$$
$$\eta_2^* = (\frac{1-i}{2}) \eta_{1,2,1}/B(\frac{1}{4}, \frac{1}{2}) \tag{2.12}$$
$$\eta_3^* = (\frac{1-i}{2}) \eta_{2,1,1}/B(\frac{1}{2}, \frac{1}{4})$$

where $B(r, s)$ is the beta function $\frac{\Gamma(r)\Gamma(s)}{\Gamma(r+s)}$. Then

$$\int_a \eta_j^* = \frac{-i}{2}, \quad j = 1, 2, 3 \tag{2.13}$$

and the other periods $\int_{K_P} \eta_j^*$ can be determined by using

$$A\eta_j^* = i\eta_j^*, \quad j = 1, 2 \qquad A\eta_3^* = -\eta_3^*$$
$$B\eta_j^* = i\eta_j^*, \quad j = 1, 3 \qquad B\eta_2^* = -\eta_2^* \tag{2.14}$$

We will take

$$\theta_1 = 2\eta_1^*, \quad \theta_2 = (1-i)(\eta_2^* - \eta_1^*), \quad \theta_3 = (1-i)(\eta_3^* - \eta_1^*). \tag{2.15}$$

Then, the periods of θ_j over the cycles (2.11) are

	K_1	K_2	K_3	K_4	K_5	K_6
$\theta_1 = 2\theta_1^*$	$-i$	1	i	$-i$	$-i$	1
$\theta_2 = (1-i)(\eta_2^* - \eta_1^*)$	0	0	$-(1+i)$	1	$1+i$	-1
$\theta_3 = (1-i)(\eta_3^* - \eta_1^*)$	0	i	0	i	$1+i$	$i-1$

Using column operation, *i.e.*, a change of basis of the K_j *over* $\mathbb{Z}(i)$, we get the new period matrix

$$\begin{matrix} \theta_1 \\ \theta_2 \\ \theta_3 \end{matrix} \begin{bmatrix} -i & 0 & 0 & 0 & 0 & 0 \\ 0 & 0 & 0 & 1 & 0 & 0 \\ 0 & i & 0 & 0 & 0 & 0 \end{bmatrix}.$$

The periods of $\theta_1 \wedge \theta_2 \wedge \theta_3 = \frac{-i}{2}(2\eta_1^* \wedge 2\eta_2^* \wedge 2\eta_3^*)$ over the last $\mathbb{Z}(i)$ basis of $H_3(J; \mathbb{Z}(i))$ clearly generate $\mathbb{Z}(i)$, and so $\theta_1 \wedge \theta_2 \wedge \theta_3$ is a $\mathbb{Z}(i)$ generator of $H_{\mathbb{Z}(i)}^{3,0}$ (holomorphic 3-forms with periods in $\mathbb{Z}(i)$). We will calculate $\nu(\theta_1 \wedge \theta_2 \wedge \theta_3) = 2I(\theta_1 \wedge \theta_2 \wedge \theta_3) \in \mathbb{C}/\mathbb{Z}(i)$. The Poincaré dual homology class $PD(\theta_2)$ is defined by $PD(\theta_2) \circ K_j = \int_{K_j} \theta_2$ ($\circ$ =intersection number) so that

$$PD(\theta_2) = (1+i)K_4 + K_3 - (1+i)K_6 - K_5. \tag{2.16}$$

We will evaluate $\tilde{I}(\theta_1 \wedge \theta_3 \wedge \theta_2) = \bar{I}(\theta_1 \wedge \theta_3 \otimes PD(\theta_2))$ as the linear combination of iterated integrals $\int_{a_j}(\theta_1, \theta_3)$, where $a_j \in \pi_1(F; 1)$ has homology class K_j, with coefficients those in (2.16). It will be simpler to write the integrand (θ_1, θ_3) as $(2\eta_1^*, (1-i)(\eta_3^* - \eta_1^*)) = 2(1-i)(\eta_1^*, \eta_3^*) - 2(1-i)(\eta_1^*, \eta_1^*)$ and recall that $2\int_\gamma(\eta_1^*, \eta_1^*) = \int_\gamma \eta_1^* \cdot \int_\gamma \eta_1^*$.

i) $a_3 = B^2a$ has base point at 1,

$$\int_{a_3}(\theta_1, \theta_3) = 2(1-i)[\int_{B^2a}(\eta_1^*, \eta_3^*) - \frac{1}{2}[\int_{B^2a}\eta_1^*]^2]$$

$$= 2(1-i)[\int_a(\eta_1^*, \eta_3^*) - \frac{1}{2}[\int_a \eta_1^*]^2]$$

(since $B\eta_1^* = -\eta_1^*, B\eta_3^* = -\eta_3^*$)

$$= 2(1-i)(\int_a(\eta_1^*, \eta_3^* + \frac{1}{8}).$$

ii) K_4 has the homology class of $A^{-1}Ba$ but $A^{-1}Ba$ has base point at $A^{-1}(1) = -i$. Let $l = (1i)_0$ be a path on sheet 0 from 1 to i, then $A^{-1}(l)^{-1}$ goes from 1 to $-i$ and we may take

$$a_4 = A^{-1}(l)^{-1} \cdot A^{-1}Ba \cdot A^{-1}(l)$$

$$\int_{a_4} (\theta_1, \theta_3) = \int_{A^{-1}Ba} (\theta_1, \theta_3) + \int_{A^{-1}(l)^{-1}} \theta_1 \cdot \int_{A^{-1}Ba} \theta_3$$

$$- \int_{A^{-1}(l)^{-1}} \theta_3 \cdot \int_{A^{-1}Ba} \theta_1$$

$$= 2(1-i)[\int_{A^{-1}Ba} (\eta_1^*, \eta_3^*) - \frac{1}{2}(\int_{A^{-1}Ba} \eta_1^*)^2$$

$$+ \int_{A^{-1}(l)^{-1}} \eta_1^* \int_{A^{-1}Ba} \eta_3^* - \int_{A^{-1}(l)^{-1}} \eta_3^* \int_{A^{-1}Ba} \eta_1^*]$$

$$= 2(1-i)[-i \int_a (\eta_1^*, \eta_3^*) + \frac{1}{8} + \int_l \eta_1^* \int_a \eta_3^* - \int_l \eta_3^* \int_a \eta_1^*]$$

To calculate $\int_l \eta_j^*$, note that $a = B(l) \cdot l^{-1}$ so that

$$-\frac{i}{2} = \int_a \eta_j^* = \int_{B(l) \cdot l^{-1}} \eta_j^* = \int_l (B-1)\eta_j^* \text{ and } \int_l \eta_1^* = \int_l \eta_3^* = \frac{-1+i}{4}, \int_l \eta_2^* = \frac{i}{4}$$

Finally, $\int_{a_4} (\theta_1, \theta_3) = 2(1-i)[-i \int_a (\eta_1^*, \eta_3^*) + \frac{1}{8}]$.

iii) $K_5 = A^2 B^2 a$, but $A^2 B^2 a$ has base point at $A^2(1) = -1$, so

$$a_5 = l \cdot A(l) \cdot A^2 B^2 a \cdot (l \cdot A(l))^{-1}$$

$$\int_{a_5} (\theta_1, \theta_3) = 2(1-i)[\int_{A^2 B^2 a} (\eta_1^*, \eta_3^*) - \frac{1}{2}(\int_{A^2 B^2 a} \eta_1^*)^2$$

$$+ \int_{l \cdot A(l)} \eta_1^* \int_{A^2 B^2 a} \eta_3^* - \int_{l \cdot A(l)} \eta_3^* \int_{A^2 B^2 a} \eta_1^*]$$

$$= 2(1-i)[- \int_a (\eta_1^*, \eta_3^*) + \frac{1}{8} + (1+i)\frac{(-1+i)}{4}(-1)(-\frac{i}{2})]$$

$$= -2(1-i) \int_a (\eta_1^*, \eta_3^*) + (\frac{-1-3i}{4}).$$

iv)

$$K_6 = A^2 B^{-1} a$$

$$a_6 = l \cdot A(l) \cdot A^2 B^{-1} a \cdot (l \cdot A(l))^{-1}$$

$$\int_{a_6} (\theta_1, \theta_3) = 2(1-i) \Big[\int_{A^2 B^{-1} a} (\eta_1^*, \eta_3^*) - \frac{1}{2} \Big(\int_{A^2 B^{-1} a} \eta_1^* \Big)^2$$

$$+ \int_{l \cdot A(l)} \eta_1^* \int_{A^2 B^{-1} a} \eta_3^* - \int_{l \cdot A(l)} \eta_3^* \cdot \int_{A^2 B^{-1} a} \eta_1^* \Big]$$

$$= 2(1-i) \Big[\int_a (\eta_1^*, \eta_3^*) - \frac{1}{8} + (1+i)\Big(\frac{-1+i}{4}\Big)(-i)\Big(\frac{-i}{2}\Big) \Big]$$

$$= 2(1-i) \int_a (\eta_1^*, \eta_3^*) + \frac{1+i}{4}.$$

Taking the linear combination given by (2.16),

$$\tilde{I}(\theta_1 \wedge \theta_3 \wedge \theta_2) = 2(1-i) \int_a (\eta_1^*, \eta_3^*)[1 + (1+i)(-i) + 1 - (1+i)]$$

$$+ \frac{1-i}{4} + \frac{1}{2} + \frac{1+3i}{4} - \frac{i}{2}$$

$$= 2(1-i)(2-2i) \int_a (\eta_1^*, \eta_3^*) \mod \mathbb{Z}(i)$$

$$= -8i \int_a (\eta_1^*, \eta_3^*) \mod \mathbb{Z}(i),$$

$$\nu(\theta_1 \wedge \theta_3 \wedge \theta_2) = -16i \int_a (\eta_1^*, \eta_3^*) \mod \mathbb{Z}(i).$$

To calculate $\int_a (\eta_1^*, \eta_3^*)$ we will consider first the automorphism σ (of order 3) of $F : x^4 + y^4 = 1$.

$$\sigma(x, y) = \Big(\frac{y}{x\sqrt{i}}, \frac{1}{ix} \Big) \quad (\sqrt{i} = \frac{1+i}{\sqrt{2}})$$

$$\sigma(\eta_3^*) = \eta_2^*, \ \sigma(\eta_2^*) = -i\eta_1^*, \ \sigma(\eta_1^*) = i\eta_3^*$$

$$\sigma(a) = A^{-1} B^2 [(01)_1 \cdot a \cdot (01)_1^{-1}]$$

$((01)_1$ is the path on sheet 1 from $x = 0$ to $x = 1$.)

$$\sigma[(1i)_1] = (01)_3 (10)_2 = B^3 [(01)_0 \cdot B^{-1} (01)_0^{-1}] = \sigma B(l),$$

$$\int_{\sigma B(l)} \eta_j^* = \int_l B^* \sigma^* \eta_j^* = \int_{B^3 (01)_0 \cdot B^2 (01)_0^{-1}} \eta_j^* = \int_{(01)_0} (B^3 - B^2) \eta_j^*$$

(recall: $l = (1i)_0$). Therefore,

$$(i^3 - i^2) \int_{(01)_0} \eta_1^* = \int_l i(i\eta_3^*) = \frac{1-i}{4}$$

or $\int_{(01)_0} \eta_1^* = \frac{1}{4}$. Similarly, $\int_{(01)_0} \eta_2^* = \frac{1-i}{8} = \int_{(01)_0} \eta_3^*$. Now

$$\int_a (\eta_1^*, \eta_3^*) = \int_{B(l)\cdot l^{-1}} (\eta_1^*, \eta_3^*)$$

$$= \int_{B(l)} (\eta_1^*, \eta_3^*) + \int_{l^{-1}} (\eta_1^*, \eta_3^*) + \int_{B(l)} \eta_1^* \int_{l^{-1}} \eta_3^*$$

$$\left(\text{ using (2.2) which implies } \int_l (\alpha, \beta) + \int_{l^{-1}} (\alpha, \beta) + \int_l \alpha \int_{l^{-1}} \beta = 0 \right)$$

$$= -\int_l (\eta_1^*, \eta_3^*) - \int_l (\eta_1^*, \eta_3^*) + \int_l \eta_1^* \int_l \eta_3^* + \int_{B(l)} \eta_1^* \int_{l^{-1}} \eta_3^*$$

$$= -2 \int_l (\eta_1^*, \eta_3^*) + (1-i) \int_l \eta_1^* \int_l \eta_3^*$$

$$= 2 \int_{B(l)} (\eta_1^*, \eta_3^*) + (1-i) \int_l \eta_1^* \int_l \eta_3^*$$

$$= 2 \int_{B(l)} (i\sigma\eta_2^*, i^{-1}\sigma\eta_1^*) + (1-i)(\frac{-1+i}{4})^2$$

$$= 2 \int_{\sigma B(l)} (\eta_2^*, \eta_1^*) - (\frac{1+i}{8})$$

$$= \int_{B^3(01)_0 \cdot B^2(01)_0^{-1}} (\eta_2^*, \eta_1^*) - (\frac{1+i}{8})$$

$$= 2[\int_{B^3(01)_0} (\eta_2^*, \eta_1^*) + \int_{B^2(01)_0^{-1}} (\eta_2^*, \eta_1^*)$$

$$+ \int_{B^3(01)_0} \eta_2^* \int_{B^2(01)_0^{-1}} \eta_1^*] - (\frac{1+i}{8})$$

$$= 2[i \int_{(01)_0} (\eta_2^*, \eta_1^*) - \int_{(01)_0} (B^2\eta_2^*, B^2\eta_1^*)$$

$$+ \int_{(01)_0} B^2\eta_2^* \int_{(01)_0} B^2\eta_1^* + \int_{(01)_0} B^3\eta_2^* \int_{(01)_0} (-B^2)\eta_1^*] - (\frac{1+i}{8}).$$

Finally, $\int_a (\eta_1^*, \eta_3^*) = 2(1+i) \int_{(01)_0} (\eta_2^*, \eta_1^*) - \frac{1}{4}$. Recall:

$$\eta_2^* = \frac{1-i}{2} \frac{1}{B(1/4, 1/2)} \frac{dx}{(1-x^4)^{1/2}}$$

(where on sheet 0, $(1 - x^4)^{1/2} = 1$ when $x = 0$).

$$\eta_1^* = \frac{1}{B(1/4, 1/4)} \frac{dx}{(1 - x^4)^{3/4}}$$

$$\int_a (\eta_1^*, \eta_3^*) = \frac{2}{B(1/2, 1/4)B(1/4, 1/4)} \int_0^1 [\int_0^x \frac{dt}{(1 - t^4)^{1/2}}] \frac{dx}{(1 - x^4)^{3/4}} - \frac{1}{4}.$$

Using: $B(1/2, 1/4)B(1/4, 1/4) = \dfrac{\Gamma(\frac{1}{4})^4}{\pi\sqrt{2}} = \dfrac{(3.6256099082)^4}{\pi\sqrt{2}}$ and

$$\int_0^1 [\int_0^x \frac{dt}{(1 - t^4)^{1/2}}] \frac{dx}{(1 - x^4)^{3/4}} = 1.5113980 \pm .007,$$

we get

$$\int_a (\eta_1^*, \eta_3^*) = 2(.03886151) \pm .0036 - 1/4$$

$$\nu(\theta_1 \wedge \theta_3 \wedge \theta_2) = 2\tilde{I}(\theta_1 \wedge \theta_3 \wedge \theta_2) = -16i \int_a (\eta_1^*, \eta_3^*)$$
$$= [1.2435683 \pm .00576] \times (-i) \mod \mathbb{Z}(i).$$

3. Some further details on

$$\int_0^1 [\int_0^x \frac{dt}{\sqrt{1 - t^4}}] \frac{dx}{(1 - x^4)^{3/4}} = \int_{(01)_0} (\eta_2, \eta_1).$$

First, the elliptic integral $\int_0^x \frac{dt}{\sqrt{1-t^4}}$ ($0 \le x \le 1$) trnasforms under the change of variable

$$\sin^2\theta = \frac{2t^2}{1 + t^2}, \quad t^2 = \frac{\sin^2\theta}{1 + \cos^2\theta}, \quad 1 - t^4 = \frac{4\cos^2\theta}{(1 + \cos^2\theta)^2}$$

$$\frac{dt}{\sqrt{1 - t^4}} = \frac{d\theta}{\sqrt{2 - \sin^2\theta}} = \frac{1}{\sqrt{2}} \frac{d\theta}{\sqrt{1 - 1/2\sin^2\theta}} = \frac{d\theta}{\sqrt{1 + \cos^2\theta}}$$

into the integral (where $x = \frac{\sin\varphi}{\sqrt{1+\cos^2\varphi}}$, $\sin^2\varphi = \frac{2x^2}{1+x^2}$)

$$\frac{1}{\sqrt{2}} \int_0^\varphi \frac{d\theta}{\sqrt{1 - 1/2\sin^2\theta}} = \frac{1}{\sqrt{2}} F(\varphi/\frac{\pi}{4})$$

(notation of Abramowitz and Steigun Bureau of Standards volume) ($0 \leq \varphi \leq \pi/2$ corresponds to $0 \leq x \leq 1$). The iterated integral is then

$$\int_0^1 [\int_0^x \frac{dt}{\sqrt{1-t^4}}] \frac{dx}{(1-x^4)^{3/4}} = \frac{1}{2} \int_0^{\pi/2} \frac{d\varphi}{\sqrt{\cos\varphi}} F(\varphi/\frac{\pi}{4}).$$

(Since $\frac{dx}{(1-x^4)^{3/4}}$ under $x = \frac{\sin\varphi}{\sqrt{1+\cos^2\varphi}}$, becomes

$$\frac{dx}{(1-x^4)^{1/2}(1-x^4)^{1/4}} = \frac{d\varphi}{\sqrt{1+\cos^2\varphi}} \frac{1}{\sqrt{2}} \frac{\sqrt{1+\cos^2\varphi}}{\sqrt{\cos\varphi}} = \frac{1}{\sqrt{2}} \frac{d\varphi}{\sqrt{\cos\varphi}}.]$$

Let $\psi = \pi/2 - \varphi$ then the integral is

$$\frac{1}{2} \int_0^{\pi/2} \frac{d\psi}{\sqrt{\sin\psi}} F(\frac{\pi}{2} - \psi/\frac{\pi}{4}).$$

To take care of the singularity of $\frac{1}{\sqrt{\sin\psi}}$ at $\psi = 0$, write

$$\frac{1}{\sqrt{\sin\psi}} = \frac{1}{\sqrt{\psi}} + \frac{1}{12}\psi^{3/2} + \frac{1}{160}\psi^{7/2} + \frac{61}{7 \times 72 \times 240}\psi^{11/2} + \cdots.$$

Since starting with

$$\frac{\sin\psi}{\psi} = 1 - \frac{\psi^2}{3!} + \frac{\psi^4}{5!} - \cdots$$

$$= (1 + a_1\psi + a_2\psi^2 + \cdots)^2,$$

we find

$$\sqrt{\frac{\sin\psi}{\psi}} = 1 - \frac{\psi^2}{2 \cdot 3!} + \frac{\psi^4}{2 \cdot 6!} - \frac{5\psi^6}{24 \cdot 7!} + \cdots$$

$$\sqrt{\frac{\psi}{\sin\psi}} = 1 + \frac{1}{2 \cdot 3!}\psi^2 + \frac{1}{160}\psi^4 + \frac{61}{7 \cdot 72 \cdot 240}\psi^6 + \cdots$$

$$\frac{1}{\sqrt{\sin\psi}} - \frac{1}{\sqrt{\psi}} - \frac{\psi^{3/2}}{12} - \frac{\psi^{7/2}}{160} = O(\psi^{11/2})$$

and is five times differentiable.

We write our integral as a sum of two terms:

$$\frac{1}{2}\int_0^{\pi/2} d\,\psi\,\left[\frac{1}{\sqrt{\sin\psi}} - \frac{1}{\sqrt{\psi}} - \frac{\psi^{3/2}}{12} - \frac{\psi^{7/2}}{160}\right]F(\frac{\pi}{2} - \psi/\frac{\pi}{4})$$

$$+ \frac{1}{2}\int_0^{\pi/2}\left[\frac{1}{\sqrt{\psi}} + \frac{\psi^{3/2}}{12} + \frac{\psi^{7/2}}{160}\right]]F(\frac{\pi}{2} - \psi/\frac{\pi}{4})d\psi$$

and integrate by parts in the second integral, noting that the integral of the first factor vanishes at $\psi = 0$, and the derivative of the second factor is $\frac{-1}{\sqrt{1-1/2\sin^2(\pi/2-\psi)}}$ while $]F(\frac{\pi}{2} - \psi/\frac{\pi}{4})$ vanishes at $\psi = \pi/2$. Thus the second integral is

$$\frac{1}{2}\int_0^{\pi/2}\left[2\psi^{1/2} + \frac{\psi^{5/2}}{30} + \frac{\psi^{9/2}}{720}\right]\frac{1}{\sqrt{1-(\cos^2\psi)/2}}d\psi$$

and with the substitution $\psi = t^2$, we get

$$\int_0^{\sqrt{\pi/2}}\left[2t^2 + \frac{t^6}{30} + \frac{t^{10}}{720}\right]\frac{1}{\sqrt{1-1/2\cos^2(t^2)}}dt.$$

Numerically (using 20 subdivisions and Simpson's rule), we get $3.0186634/2 = 1.5093$ for this integral and for the first integral, get $.0035718/2$. Thus for $\int_{(01)_0}(\eta_2, \eta_1)$ we get 1.5111 and for $\frac{2}{B(1/2,1/4)B(1/4,1/4)}\int_0^1[\int_0^x \frac{dt}{(1-t^4)^{1/2}}]\frac{dx}{(1-x^4)^{3/4}}$ we get the value given before.

[1] G. Ceresa, "C is not algebraically equivalent to C^- in its Jacobian, " Annals of Math. vol. 117, 1983.

[2] P. Griffiths, "On the periods of certain rational integrals II, " Annals of Math. vol. 90, 1969, pp. 496-541.

[3] P. Griffiths, "Some transcendental methods in the study of algebraic cycles," Lecture Notes in Math., vol. 185, pp. 3-4, Springer-Verlag, Berlin, 1971.

[4] B. Gross, D. Rohrlich, "On the periods of abelian integrals and a formula of Chowla and Selberg, " Inv. math. vol. 45, 1978, pp. 193-211.

[5] B. Harris,"'Harmonic Volumes," Acta Mathematica, vol. 150, 1983 pp. 91-123.

[6] A. Weil, Collected Works, vol. 2, pp. 533-534.

2.8   Currents and Hodge theory

In order to generalize to higher dimensions some of the calculations we have carried out on Riemann surfaces, we have to replace the technique of cutting up these 2-manifolds along curves and considering functions with jumps across the cuts by more general methods suited to higher dimensional manifolds and submanifolds. As preparation for doing this in the next chapter we review here some of the theory of currents and associated Hodge and de Rham theory.

Let X be a compact C^∞ manifold of dimension n, oriented, without boundary. $A^p(X)$=space of C^∞ p-forms (real valued).

Definition 2.2   A current T on X of dimension p (or degree $n-p$) is an $\mathbb{R}$-linear function $T : A^p(X) \to \mathbb{R}$ satisfying a continuity condition: there exists an integer k and a finite number of coordinate neighborhoods U_j covering X such that if in U_j we write a p-form ϕ as $\phi = \sum \phi_I(x)dx^I$ (I=multi-index) then for some constant C,

$$|T(\phi)| \leq C \sum_I \sum_{|J| \leq k} \sup |\partial^J \phi_i|$$

(J again denotes multi indices, ∂^J=partial derivative). We also say T is of order $\leq k$.

Example 2.1   Let τ be a C^∞ $n-p$ form on X and let $T(\phi) = \int_X \phi \wedge \tau$. Here $k = 0$.

Example 2.2 S is an oriented C^∞ submanifold of X of dimension p, and

$$T(\phi) = \int_S \phi.$$

Here we write $T = \delta_S =$ "Dirac current given by S". We allow S to have a boundary.

We can generalize this example in two ways: instead of having an embedding $S \subset X$, we can just be given a smooth map $f : S \to X$ and let

$T(\phi) = \int_S f^*\phi$. Further we can take several such S: say $S_1, \cdots, S_n$ and real coefficients $c_1, \cdots, c_n$ and define $T(\phi) = \sum_{i=1}^n c_i \int_{S_i} f_i^*\phi$. We write $S = \sum c_i(S_i, f_i)$, $T = \delta_S$.

Example 2.3   Let $X = S^2 = \mathbb{R}^2 \cup \infty$, $g(x, y) =$ any L^1 function (say on $\mathbb{R}^2$), e.g, $\log \frac{|z|^2}{1+|z|^2}$. $p = 2$ and $T(\phi) = \int_{\mathbb{R}^2} g(x, y)\phi$.

Recall that T is said to have dimension p and degree $n - p$. We now define the exterior differential dT as the current of degree $p + 1$ given by (for any $p - 1$ form β)

$$dT(\beta) = (-1)^p T(d\beta).$$

If we use the *notation*

$$T(\phi) = \int_X \phi \wedge T = (\phi, T)$$

then $\int_X \beta \wedge dT$ is defined as $(-1)^p \int_X d\beta \wedge T$ in accordance with example 2.1 where T is given by an $n - p$ form τ.

In example 2.2, $T = \delta_S$ where S has boundary ∂S, we get $dT = (-1)^p \delta_{\partial S}$. Next we look at dT for T given in example 3, where $X = \mathbb{R}^2 \cup \infty = S^2$ and T is given by its value on a 2-form ϕ on S^2 as

$$T(\phi) = \int_{S^2} \phi \log(\frac{r^2}{1 + r^2})$$

$$= \lim_{\epsilon \to 0} \int_{r \geq \epsilon} \phi \log(\frac{r^2}{1 + r^2}).$$

Here $dT = d\log(\frac{r^2}{1+r^2})$, as current of degree 1.

If X is a complex manifold with complex structure operator J (which acts on tangent vectors and on differential forms), we define $J(T)$ for a current T so as to agree with the definition of $J(\tau)$ for a form τ if T is given by τ: since we have

$$\int_X J(\phi) \wedge J(\tau) = \int_X \phi \wedge \tau$$

$(J(\phi \wedge \tau) = J(\phi) \wedge J(\tau))$, we define

$$(\phi, J(T)) = (J^{-1}(\phi), T).$$

As example we consider S^2 with polar coordinates r, θ. Then

$$J(dz/z) = idz/z, \quad J(dr/r) = -d\theta.$$

If T is the current given by $\log(r^2)$, then $J^{-1}(dT) = 2d\theta$.

The operator d^c is defined as

$$d^c = \frac{1}{4\pi} J^{-1} \circ d \circ J = (i/4\pi)(\bar\partial - \partial)$$

so $dd^c = (i/2\pi)\partial\bar\partial$.

If $T = \log r^2$ then

$$d^c \log r^2 = (1/2\pi)d\theta$$
$$dd^c \log r^2 = d(d\theta/2\pi) = \delta_0 - \delta_\infty.$$

Thus, for a differentiable function $\phi(x, y)$ on $\mathbb{S}^2$,

$$(\phi, d(d\theta/2\pi)) = -\int_{\mathbb{S}^2} d\phi \wedge (d\theta/2\pi) = \phi(0) - \phi(\infty).$$

We generalize the last example as follows:

1. On $\mathbb{S}^2 = \mathbb{P}^1(\mathbb{C})$ we have the complex line bundle $\mathcal{O}(1)$ whose fiber over each 1-(complex) dimensional linear subspace $\mathbb{C}(z_0, z_1)$ of $\mathbb{C}^2$ is the dual space $\mathbb{C}(z_0, z_1)^*$. We have a hermitian metric on each fiber, given by: if $l : \mathbb{C}(z_0, z_1) \to \mathbb{C}$ is linear then

$$||l||^2 = \frac{|l(z_0, z_1)|^2}{|(z_0, z_1)|^2} = \frac{|l(z_0, z_1)|^2}{z_0\bar{z}_0 + z_1\bar{z}_1}.$$

Let now s be the holomorphic section of $\mathcal{O}(1)$ such that $s(\mathbb{C}(z_0, z_1))=$linear function l, $l(z_0, z_1) = z_1$. Then $||s||^2$ is a function on $\mathbb{P}^1$ given by $||s||^2 = |z_1|^2/(|z_0|^2 + |z_1|^2)$ and: $dd^c \log ||s||^2 = \delta_0 - \omega$ where ω is a 2-form on $\mathbb{P}^1$ invariant under the unitary group $U(2)$ and satisfying $\int_{\mathbb{P}^1} \omega = 1$. ω is the (first) Chern form of $\mathcal{O}(1)$. Finally z_1/z_0 is a meromorphic function on $\mathbb{P}^1$ with divisor $div(z_1/z_0) = (0) - (\infty)$. Write $z = z_1/z_0$: then we have an equation involving currents:

$$dd^c \log |z|^2 = \delta div(z_1/z_0).$$

The following is a general fact, called the Poincaré-Lelong formula (see [Griffiths-Harris]): for any complex analytic line bundle L over a complex manifold X with hermitian metric on the fibers and s a non-zero section of L (holomorphic or meromorphic) with divisor $div\ s = Z = Z_0 - Z_\infty$, we have

$$dd^c \log ||s||^2 = \delta_{z_0} - \delta_{z_\infty} - \omega = \delta_{div\ s} - \omega,$$

where ω is the Chern form defined so that it is a $(1, 1)$ form representing the Poincaré dual cohomology class to the homology class of *div s*.

De Rham's results for currents:

The first result is that the cohomology of the space of currents for the operator d is the same as the cohomology of the subspace of differential forms: that is, the inclusion of $A^*(X)$ into the space of all currents (for compact oriented X) induces a cohomology isomorphism. In particular, if α is a closed form on X and $\alpha = dT$ for a current T then $\alpha = d\tau$ for a form τ.

As an application, let S be a compact oriented submanifold of X, of dimension p, *without boundary*, (or more generally a linear combination of pairs $(S_i, f_i : S_i \rightarrow X)$ as before). Then δ_S is a current and $d(\delta_S) = 0$. Let ω_S be a closed form on X which represents the Poincaré dual cohomology class to the homology class of S (in singular homology of X). Considering the cohomology of currents on X, both δ_S and ω_S represent the same cohomology class, and so their difference is exact:

$$\delta_S - \omega_S = dT$$

where T is a current of degree $n - p - 1$ (or dimension $p + 1$). However we will want to say more about being able to choose a T with desirable properties.

Next we consider X as above together with a Riemannian metric. We then have operators d and d^*, $\Delta = dd^* + d^*d$ on currents as well as on forms (we define $\star$, d^*, in fact all operators on currents so as to agree with the definition of these operators on the subspace of differential forms. However currents are not given a pre-Hilbert space structure). The Hodge decomposition of $A^*(X)$ into the three subspaces $\mathrm{Im}d$, $\mathrm{Im}d^*$, $\mathcal{H}$, which followed from the fact that if α is orthogonal to $\mathcal{H}$ then there is a unique β with $\Delta\beta = \alpha$, in other words $A^*(X) = \mathcal{H} \oplus \Delta A^*(X)$, Δ an isomorphism on $\mathcal{H}^\perp$, now says that if a current T annihilates $\mathcal{H}$ then it satisfies $T = \Delta S$ for a unique current S which annihilates $\mathcal{H}$. This gives a decomposition of the space of currents into the same kinds of subspaces: $\mathrm{Im}d$, $\mathrm{Im}d^*$, $\ker \Delta$.

We now wee that a harmonic current has to be equal to a harmonic form considered as a current (since both map isomorphically to cohomology). However a stronger result holds: *The regularity theorem for the Laplacian* $\Delta = dd^* + d^*d$. To formulate this we have to define the notion of smoothness of a current T (on X) on an open subset U of X. For non-compact manifolds such as U, one considers forms $A_c^*(U)$ with compact support in U as a subspace of $A^*(X)$, and defines currents on U in a similar way to current on X. Then currents on X restrict to (currents on) U. We say that a current T on X is smooth on U if there is a smooth form τ_U on U (not necessarily with compact support on U) such that for all $\phi \in A_c^*(U)$, $T(\phi) = \int_X \phi \wedge \tau_U (= \int_U \phi \wedge \tau_U)$. The regularity theorem now states that if S, T are currents on X and $\Delta S = T$ then for any open U, if T is smooth on U then S is also smooth on U.

In particular if T is smooth on X then so is S, and if $T = 0$ on X then S is smooth and harmonic on X. Another consequence of regularity is the following: as before let δ_S be the Dirac current of a cycle and ω_S a closed form representing the Poincaré dual cohomology class. Suppose T is a current such that

$$dT = \delta_S - \omega_S$$

and furthermore T *is coexact* $(T = d^*T')$. Then: 1. T is uniquely determined by δ_S and ω_S, and

2. T is smooth outside the support of S (*i.e.*, outside the union of the images $f_i(S_i)$ if $S = \sum a_i(S_i, f_i)$). To see this, note that if $T = d^*T'$ then we may assume $T' \in \mathrm{Im}d$, so $T = d^*dT_1$ and so

$$dT = dd^*(dT_1) = (dd^* + d^*d)(dT_1) = \Delta(dT_1).$$

Thus $\Delta(dT_1) = \delta_S - \omega_S$ is smooth outside the support S, so dT_1 is smooth on this set and finally $d^*(dT_1) = T$ is also smooth there.

Finally, we may choose ω_S to be harmonic and then T, assumed coexact, is uniquely determined by S and by the metric on X and is coexact outside S.

If we do not fix a metric, then, given S, T exists but is not unique. The situation is better on a complex X with Kahler metric: we will use both d and d^c there and the dd^c lemma.

The regularity theorem on a Kahler X is of course the same, but now the Laplacian Δ is

$$\Delta = dd^* + d^*d = 16\pi^2(d^c d^{c*} + d^{c*}d^c)$$

and the space of currents is the direct sum of $\mathcal{H}$ and 4 other subspaces, images of $dd^c, d^*d^c, d^{c*}d, d^{c*}d^*$. A regularity theorem for the operator dd^c is:

If a current T is dd^c exact and T is smooth on U then $T = dd^cT_1$ for a unique current $T_1 \in$ Image of d^*d^{c*}, and T_1 is smooth on U. The proof is the same as in the Riemannian case.

To sum up, the regularity theorem on a Riemannian (compact) X implies all the other facts stated above.

In the next chapter we will use currents T which are given by multiplying by an L^1 form τ and integrating over the manifold X: that is τ will be continuous on the complement of a submanifold S of lower dimension and, if τ has degree p and α is any $n - p$ form continuous on all of X then we require that $\alpha \wedge \tau$ be an absolutely integrable n-form and $T(\alpha) = \int_X \alpha \wedge \tau$. We will say for short that T is given by an L^1 form. We will also use Dirac δ type currents δ_S which are not L^1. The main example of an L^1 current will be the angular form around a submanifold S, gerenalizing $d\theta / 2\pi$ around the origin in the plane.

Chapter 3

The Generalized Linking Pairing and the Heat Kernel

3.1 Introduction

In this chapter we will generalize some of the results of chapter 2 to higher dimensional Riemannian manifolds (always assumed compact oriented even dimensional); for compact complex Kahler manifolds our invariants will be independent of the choice of Kahler metric and depend only on the complex structure.

We will begin with a Riemannian manifold X (as above) and a construction assigning to a pair of cycles A, B satisfying two further conditions, a real number (A, B) which is a generalization of the ordinary topological linking number (an integer). A, B are assumed to have dimensions p, q satisfying:

$$p + q = dimX - 1 \tag{3.1}$$

Further, A and B must have disjoint supports, thus each of A, B is a formal linear combination (with integer coefficients) of differentiable maps of a standard simplex to X, and the support is the union of the images of these maps.

The real number (A, B) is obtained as follows: if Δ is the Laplacian on forms on X, then the heat operator $\exp(-t\Delta)$ has a kernel $K(x, y, t)$, an n-form on $X \times X$ for $n = dimX$ (for each $t > 0$), and we construct a related form $\Gamma(x, y, t)$ of degree $n-1$ and integrate Γ over $A \times B$, then take the limit as $t \to 0$. If A, B are both homologous to 0 in X then (A, B) is an integer, the ordinary linking number (which is independent of the metric). If just A bounds, it follows that (A, B) *mod* $\mathbb{Z}$ depends only on the homology class of B and (A, B) *mod* $\mathbb{Z}$ as function of the homology class of B, is just the Abel-Jacobi image of A in a torus. If $A = \partial C$ and ω_B is the harmonic form Poincare dual to (the homology class of) B, then (A, B) *mod* $\mathbb{Z}$ is given by $-\int_C \omega_B$ (*mod* $\mathbb{Z}$).

We will use the pairing (A, B) also in situations where neither of A, B is homologous to 0. See [Harris, 1993].

Let X have dimension $n = 2m$, and consider the heat operator $\exp(-t\Delta)$ for all real $t > 0$, acting on forms α on X. Our main reference will be the book *Heat Kernels and Dirac Operators* by Berline, Getzler, and Vergne [BGV]. See also our paper [Harris, 1993]. The operator $\exp(-t\Delta)$ is given by a kernel $K(x, y, t)$ which is a C^∞ n-form on $X \times X \times (\mathbb{R} > 0)$:

$$[\exp(-t\Delta)\alpha](y) = \int_{x \in X} \alpha(x) \wedge K(x, y, t) = pr_{2*}(pr_1^* \alpha \wedge K) \qquad (3.2)$$

in the notation of the appendix on orientations and fiber integration. As $t \to 0$, this expression approaches $\alpha(y)$.

$K(x, y, t)$ is given by an infinite series if we choose an orthonormal basis of the forms $A^*(X)$ which are eigenforms of the Laplacian Δ of X: denoting this basis α_λ^p with $\Delta\alpha_\lambda^p = \lambda\alpha_\lambda^p$ and ordering the eigenvalues (for each degree p) as $0 \le \lambda_1 \le \lambda_2 \le \dots$ we get a series expansion:

$$K(x, y, t) = \sum_p \sum_\lambda \exp(-\lambda t)(\star\alpha_\lambda^p(x)) \wedge \alpha_\lambda^p(y) \qquad (3.3)$$

where $x, y \in X$, $t > 0$, which is uniformly convergent, as are all x, y, t derivatives, because the eigenvalues tend sufficiently rapidly to ∞. Thus we can integrate term-by-term or similarly differentiate without worrying about convergence with respect to any of the variables x, y, t (for $t > 0$).

By orthonormality of the (real) forms α_λ^p we mean $\int_X \alpha_\lambda^p \wedge \star\alpha_\mu^q = 0$ if $\alpha_\lambda^p \ne \alpha_\mu^q$ and $= 1$ otherwise. The forms with $\lambda = 0$ are a basis for the harmonic forms and the corresponding part of the series, denoted

$$H(x, y) = \sum_p \sum_{\lambda=0} \star\alpha_0^p(x) \wedge \alpha_0^p(y) \qquad (3.4)$$

is the kernel for the projection onto harmonic forms. Also, $H(x, y) =$ limit of $K(x, y, t)$ as $t \to \infty$. We will use d or d^* to denote operators in the main variables (x, y).

We can either use the series or appeal to the fact that $\exp(-t\Delta)$ commutes with exterior differential d to see that $K(x, y, t)$ is a closed form (for each t) on $X \times X$; also we can check that its DeRham cohomology class on $X \times X$ is Poincare dual to the diagonal. H is clearly a closed form as well, and Poincare dual to the diagonal.

On $X \times X$ we use the product metric: then the Laplacian Δ on $X \times X$ is the sum $\Delta = \Delta_1 + \Delta_2$ of the Laplacians on the individual factors. $K(x, y, t)$ satisfies a heat equation :

$$\Delta K(x, y, t) = -2\frac{\partial K}{\partial t} \tag{3.5}$$

We can now give a formula for the coexact (on $X \times X$) form $\Gamma(x, y, t)$ satisfying

$$d\Gamma(x, y, t) = K(x, y, t) - H(x, y) \tag{3.6}$$

Theorem 3.1

$$\Gamma(x, y, t) = \frac{1}{2}\int_{\tau=t}^{\infty}[d^*K(x, y, \tau)]d\tau \tag{3.7}$$

where d^* is the adjoint of d on $X \times X$.

Proof: Write $K(x, y, t) - H(x, y) = K(x, y, t) - K(x, y, \infty)$

$= \int_t^{\infty}(-\frac{\partial K}{\partial \tau})d\tau$

$= \int_t^{\infty}\frac{1}{2}(\Delta K)d\tau$ by the heat equation for K

But $\Delta K = dd^*K$ since $dK = 0$, so the last integral above is equal to

$\frac{1}{2}\int_t^{\infty}(dd^*K)d\tau$

$= d[\int_t^{\infty}\frac{1}{2}(d^*K)d\tau].$

Putting Γ as the expression in square brackets, it is in the image of d^* and $d\Gamma = K - H$, concluding the proof. $\square$

We will discuss the behavior of $\Gamma(x, y, t)$ as t decreases to 0 in detail below, and will just say here that Γ approaches pointwise a C^∞ form on the complement of the diagonal $x = y$. This limit form is singular on the diagonal but is L^1 on $X \times X$ and Γ approaches it while satisfying Lebesgue dominated convergence. On the complement of a neighborhood of the diagonal the approach of Γ to this limit is uniform. The limit form is just an "angular form" in the normal direction to the diagonal, that is, its integral over any small normal $n - 1$ sphere approaches 1 as the radius (distance to the diagonal) approaches 0.

Once we have discussed the behavior of Γ (for $t \to 0$) on the complement of a neighborhood of the diagonal, we can take the disjoint cycles A, B on X and form $A \times B$ on $X \times X$ so that the following expression makes sense.

Definition 3.1

$$(A, B) = \lim_{t \to 0} \int_{A \times B} \Gamma(x, y, t). \tag{3.8}$$

The "angular form" $\lim_{t \to 0} \Gamma$ then explains why (A, B) is a generalization of the linking number .

We will need some estimates on $K(x, y, t)$ and $\Gamma(x, y, t)$: we may talk of these as functions rather than forms by using pointwise norms on forms given by the Riemannian metric. The main case to consider is when t is near 0 and x is close to y. We denote the square of the Riemannian distance by

$$dist(x, y)^2 = r(x, y)^2 = r^2. \tag{3.9}$$

The main fact is that for small t and r^2, $K(x, y, t)$ is approximately the Euclidean heat kernel $Q_t(x, y)$ in $\mathbb{R}^n \times \mathbb{R}^n \times \mathbb{R}_+$, given by (for $n = dimX$):

$$Q_t(x, y) = (4\pi t)^{-n/2} e^{-|x-y|^2/4t} (dy_1 - dx_1) \dots (dy_n - dx_n). \tag{3.10}$$

Appealing to [BGV] lemma 2.39 (pages 92,93) to bound the derivative $d^* K(x, y, t)$ we get:

For $0 < t \leq 1$, $d^* K(x, y, t)$ is bounded uniformly on $X \times X$ by $(const)t^{-(n+1)/2}$ (it suffices to see that if one differentiates the Euclidean $Q_t(x, y)$ with respect to x or y, one calculates that $\partial_x(e^{-x^2/t}) = O(t^{-1/2})$).

For $t \geq 1$, any x or y derivative of $K(x, y, t)$ is bounded by a constant times $e^{-t\lambda/2}$ for λ the smallest eigenvalue > 0 of Δ (BGV Proposition 2.36) and so the integral $\int_1^\infty (d^* K)dt$ is uniformly bounded on $X \times X$.

Putting these two bounds together we can write:

$$\Gamma(x, y, t) = \frac{1}{2} \int_t^\infty [d^* K(x, y, \tau)]d\tau \tag{3.11}$$

is bounded uniformly in x, y for $t \geq 1$ and bounded by $t^{(1-n)/2}$ again uniformly in (x, y).

For $K(x, y, t)$ we have bounds:

$C_1 e^{-\lambda t/2}$ for $t \geq 1$ and

$C_2(4\pi t)^{-n/2} e^{-r^2/4t}$ for $0 < t \leq 1$ where $r^2 = dist(x, y)^2$.

For $r \geq \epsilon > 0$, ϵ fixed, $K(x, y, t)$ tends to 0 faster than any power of t, as $t \to 0$.

Since these bounds are uniform on $X \times X$, they persist after we integrate in y over a compact subset C of X:

Lemma 3.1   *Define*

$$K_C(x, t) = \int_{y \in C} K(x, y, t) \tag{3.12}$$

$$\Gamma_C(x, t) = \int_{y \in C} \Gamma(x, y, t). \tag{3.13}$$

Then $\Gamma_C(x, t)$ is $O(t^{(1-n)/2})$ as $t \to 0$, and $K_C(x, t)$ is $O(t^{-n/2} e^{-\epsilon^2/4t})$ for (distance x to C) $\geq \epsilon$ and $t \to 0$.

Lemma 3.2   *Let $C_1, \ldots, C_k$ be chains in X (over which one can integrate forms) with $k \geq 3$ and suppose $C_2 \cap \ldots \cap C_k$ is empty (i.e. the intersection of these $k - 1$ supports is empty). Then*

$$\lim_{t \to 0} \int_X \Gamma_{C_1}(x, t) \wedge K_{C_2}(x, t) \wedge \ldots \wedge K_{C_k}(x, t) = 0 \tag{3.14}$$

Proof: We can find $\epsilon > 0$ so that for each $x \in X$, there is a C_i, $2 \leq i \leq k$, such that $dist(x, C_i) \geq \epsilon$. Then $K_{C_i}(x, t)$ tends to 0 like $e^{-\epsilon^2/4t}$ as $t \to 0$, and the other factors are at most t^m. So the integrand is uniformly bounded on X and rapidly convergent to 0. $\square$

We need to extend this lemma to the case where X is replaced by a cycle $C_1 \cap \ldots \cap C_{i-1}$, assuming this intersection is defined and has the correct dimension - the main case being where $C_1, \ldots, C_{i-1}$ are submanifolds intersecting transversely.

Lemma 3.3   With $C_1 \cap \ldots \cap C_{i-1}$ satisfying the condition above and $C_1 \cap \ldots \cap C_{i-1} \cap C_{i+1} \cap \ldots \cap C_k$ being empty (i.e. the support is empty) we have:

$$\lim_{t \to 0} \int_{C_1 \cap \ldots \cap C_{i-1}} \Gamma_{C_i, t} \wedge K_{C_{i+1}, t} \wedge \ldots \wedge K_{C_k, t} = 0 \qquad (3.15)$$

Proof: Again there exists $\epsilon > 0$ such that each $x \in$ (support of) $C_1 \cap \ldots \cap C_{i-1}$ has distance $\geq \epsilon$ from at least one of $C_{i+1}, \ldots, C_k$, so the previous proof applies. $\square$

We need now information on the behavior of $\Gamma(x, y, t)$ as $t \to 0$ for (x, y) in a small neighborhood of the diagonal. Following [BGV], page 82 and Theorem 2.30, write

$$r(x, y)^2 = ||\xi||^2 \qquad (3.16)$$

where $\xi \in T_y(X)$ and $x = \exp_y(\xi)$ is the image of ξ under the exponential map from T_y to a neighborhood of y. In $\mathbb{R}^n$ we could just write $\xi = x - y$. Also write

$$k_t^N(x, y) = (4\pi t)^{-n/2} \exp(-r^2/4t) \sum_{i=0}^{N} t^i \Phi_i(x, y) \qquad (3.17)$$

for x near y and $k_t^N = 0$ for (x, y) outside the neighborhood of the diagonal, (we are omitting factors $|dx|^{1/2}, |dy|^{1/2}$ used in [BGV]), where $\Phi_i(x, y)$ are C^∞ sections (everywhere) of the bundle $Hom(\Lambda^*(T_X^*), \Lambda^*(T_X^*))$ with $\Phi_0(x, x) =$ identity: we will write, using $\mathbb{R}^n \times \mathbb{R}^n$ notation,

$\Phi_0(x,y) = \sum(\star dx_I) \wedge dy_I$ modulo higher order terms in $r(x,y)$ (where I denote multi-indices $i_1 < \ldots < i_p$).

Theorem 2.30 gives the following pointwise, in (x,y), estimate: let $N = (n/2) + 1$, then

$$d^*K(x,y,t) - d^*k_t^N(x,y) = O(t^{1/2}) \qquad (3.18)$$

uniformly for $(x,y) \in X \times X$, as $t \to 0$. Recall that:

$$\Gamma(x,y,t) = \int_t^\infty \frac{1}{2} d^*K(x,y,\tau)d\tau = \int_t^1 + \int_1^\infty . \qquad (3.19)$$

The second integral defines a C^∞ function of (x,y) on $X \times X$ ([BGV] Proposition 2.37, page 93). We thus have to study $\int_t^1 d^*K(x,y,t)$, which by the above $O(t^{1/2})$ estimate can be written as $\int_t^1 [d^*k_\tau(x,y)]d\tau + O(1)$ (as $t \to 0$), (where we write k_t for k_t^N with $N = (n/2) + 1$). Thus we have to calculate:

A. The Euclidean case, which is the "initial term" of d^*k_t, i.e.

$$\int_t^1 [\frac{1}{2}d^*Q(x,y,\tau)]d\tau \qquad (3.20)$$

where $Q(x,y,t)$ is the Euclidean heat kernel

$$\begin{aligned} Q_t(x,y) = Q(x,y,t) &= (4\pi t)^{-n/2} \exp(-|x-y|^2/4t) \\ &\times d(y_1 - x_1) \wedge \ldots \wedge d(y_n - x_n). \end{aligned} \qquad (3.21)$$

B. The influence on A, from adding to Q extra terms $Q.(x_i - y_i).\varphi_i(x,y)$ or $Q.t^j.\varphi_j(x,y)$ where $j \geq 1$ and $\varphi(x,y)$ are differentiable functions.

We start with A and note that if we define

$f : \mathbb{R}^n \times \mathbb{R}^n$
$f(x,y) = y - x = \xi,$
then
$Q(x,y,t) = f^*k(\xi,t)$ where $k(\xi,t) = (4\pi t)^{-n/2} \exp(-|\xi|^2/4t)d\xi_1 \ldots d\xi_n.$

f^* does not commute with $d^* = -\star d\star$ because f^* does not induce an isometry of $T_0^*\mathbb{R}$ with the normal to the diagonal at (x,x):

$f^*(d\xi_i) = dy_i - dx_i$ has length 2, so

$$d^*_{\mathbb{R}^n \times \mathbb{R}^n} Q(x, y, t) = 2 f^* d^*_{\mathbb{R}^n} (k(\xi, t)). \tag{3.22}$$

Thus we have to calculate

$$\Gamma(x, y, t) = \int_t^1 \frac{1}{2} d^* Q(x, y, \tau) d\tau = f^* \int_t^1 d^* k(\xi, \tau) d\tau. \tag{3.23}$$

To calculate $d^* k(\xi_1, \ldots, \xi_n, t)$ we use the following notation:

$$r^2 = \sum_1^n \xi_i^2 \ , \star d(r^2) = 2r^n \tilde{d\theta} \tag{3.24}$$

where the $(n-1)$-form $\tilde{d\theta}$ denotes the unnormalized angular form around the origin of $\mathbb{R}^n$ (rotation invariant form on any sphere S^{n-1} with center 0) with

$\int_{S^{n-1}} \tilde{d\theta} = 2\pi^{n/2}/\Gamma(n/2)$

($\Gamma(n/2)$ here denotes the gamma function, not the kernel). We let

$d\theta = \Gamma(n/2)(2\pi^{n/2})^{-1} \tilde{d\theta}$

be the normalized angular form on $\mathbb{R}^n \setminus \{0\}$ ($\int_{S^{n-1}} d\theta = 1$). This form is L^1 on $\mathbb{R}^n$.

We can now calculate (helped by [Hein])

$d^* k(\xi, t) = - \star d \star [(4\pi t)^{-n/2} \exp(- \sum \xi_i^2/4t) d\xi_1 \wedge \ldots \wedge d\xi_n]$

$= - \star d[(4\pi t)^{-n/2} \exp(-r^2/4t)]$

$= 2\pi[(4\pi t)^{-n/2-1} \exp(-r^2/4t)] r^n \tilde{d\theta}$

$= 2\pi^{-n/2}[(4t)^{-n/2-1} \exp(-r^2/4t)] r^n \tilde{d\theta}.$

Fix $r^2 \neq 0$ and make the change of variable $u = r^2/4t$, $du/u = -dt/t$, so

$$d^*k(x,t)dt = -[(2\pi^{n/2})^{-1}e^{-u}u^{n/2}du/u]\tilde{d\theta}$$

$$\int_t^1 d^*k(\xi,\tau)d\tau = [\int_{v=r^2/4}^{r^2/4u} e^{-v}v^{n/2}dv/v](2\pi^{n/2})^{-1}\tilde{d\theta}$$

Note that the integrand is positive, and as $t \to 0$ with r^2 fixed, the integral increases to $\int_{r^2/4}^\infty e^{-v}v^{n/2}dv/v$. As $r \to 0$ this last integral increases to the (gamma function) $\Gamma(n/2)$. We conclude: if we pull back to $\mathbb{R}^n \times \mathbb{R}^n$ by f^*, $\int_t^1 \frac{1}{2}d^*Q(x,y,\tau)d\tau$ is bounded by the locally L^1 form $f^*(d\theta)$ and as $r \to 0$ it approaches $f^*d\theta$ (satisfying Lebesgue dominated convergence). Here $f^*d\theta$ is the angular form around the diagonal in $\mathbb{R}^n \times \mathbb{R}^n$ and has integral 1 over every S^{n-1} normal to, and centered on, the diagonal. $d\theta$ as $n-1$ form on $\mathbb{R}^n$ is (up to a constant)

$$\tfrac{1}{|\xi|^n}\sum_{i=1}^n (-1)^{i-1}\xi_i d\xi_1 \wedge \ldots \wedge d\xi_{i-1} \wedge d\xi_{i+1} \wedge \ldots d\xi_n$$

and since $\xi_i/|\xi|$ is bounded, its pointwise norm is $1/r^{n-1}$ where $r = |\xi|$, which is the norm needed so that its integral over any $(n-1)$-sphere is independent of the radius. If we were to multiply it by any positive power of r and by an everywhere continuous function, the resulting $n-1$ form when integrated over an S^{n-1} of radius r would approach 0 with r. On pulling back to $\mathbb{R}^n \times \mathbb{R}^n$ by f^* the same properties would hold in the normal direction to the diagonal. Such a multiplication of $f^*d\theta$ by r occurs in the asymptotic expansion $k_t^N(x,y)$ in the initial term $i = 0$ when $\Phi_0(x,y)$ is expanded in a Taylor series in powers of $(x_j - y_j) = \xi_j$ (using coordinates in the tangent space T_y, i.e. geodesic coordinates centered at y), since as before $(x_j - y_j)/r$ is bounded and $x_j - y_j = r \cdot (x_j - y_j)/r$.

The other terms, those involving $t^k\Phi_k(x,y)$ in the asymptotic expansion with $k > 0$, require a calculation of

$$\int_t^1 \tau^k d^*Q(x,y,\tau)d\tau$$

which is almost the same as what was just done with $k = 0$ but with $\tau^{-n/2}$ replaced by $\tau^{-n/2+k}$. The same change of variable $v = r^2/4\tau$ now gives the integral

$$[\int_{v=r^2/4}^{r^2/4t} v^{n/2-k}e^{-v}dv/v]r^{2k}d\theta$$

so we can estimate such terms as bounded by the initial Euclidean integral times r^{2k}. In conclusion, the asymptotic expansion of $K(x, y, t)$ with respect to powers of t gives for $\Gamma(x, y, t)$ a "leading term" which as $t \to 0$ is bounded above by the Euclidean angular form $f^* d\theta$ (which is locally L^1 and of order $1/r^{n-1}$ in the normal direction to the diagonal, r being the distance to the diagonal) and approaches this angular form as $r \to 0$, with the rest of the terms after the leading one being of orders $1/r^{n-k-1}$ with $k > 0$. In particular, $\Gamma(x, y, t)$ as $t \to 0$ approaches an L^1 form on $X \times X \backslash$ (diagonal) defining a current $\Gamma(x, y)$ that satisfies

$$d\Gamma(x, y) = \delta_\Delta - \omega_\Delta$$

on $X \times X$, $\Delta = $ diagonal, $\delta_\Delta = $ Dirac current, $\omega_\Delta = $ harmonic Poincare dual form to the diagonal, $\Gamma(x, y) = $ "angular current" .

We want to see now that if the angular current $\Gamma(x, y)$ for $\Delta \subset X \times X$ is restricted to $X \times C$ for an oriented submanifold C of X and then integrated over $y \in C$, one obtains an angular form for $C \subset X$, denoted Γ_C.

Using the asymptotic behavior argument, we are reduced to looking at $X = \mathbb{R}^n$, n even, $C = \mathbb{R}^{n_2}$ where $n = n_1 + n_2$ and $\mathbb{R}^n = \mathbb{R}^{n_1} \oplus \mathbb{R}^{n_2}$ is written as $X = C^\perp \oplus C$ (everything being in a neighborhood of the origin in $\mathbb{R}^n$).

$C^\perp$ has coordinates $(x_1, \ldots, x_{n_1}, 0, \ldots, 0)$ and normal bundle orientation given by $\omega_{C^\perp} = dx_{n_1+1} \wedge \ldots \wedge dx_n$ (see appendix on orientations).

C has the usual tangential orientation $\tau_C = \omega_{C^\perp}$ and normal orientation $\omega_C = (-1)^{n_1} dx_1 \wedge \ldots \wedge dx_{n_1} = (-1)^{n_1} \tau_{C^\perp}$ (so $\tau_C \wedge \omega_C = dx_1 \wedge \ldots \wedge dx_n$).

The angular form around C, $d\theta_C$ has, as current, differential $d(d\theta_C) = \delta_C = $ limit of Gaussian forms (Gaussian function of r_1, t) $\times \omega_C$, where $r_1^2 = x_1^2 + \ldots x_{n_1}^2 = $ square of distance to C. Thus $d\theta_C$ when restricted to $C^\perp = \mathbb{R}^{n_1}$ is $(-1)^{n_1} d\theta_1$, $d\theta_1$ denoting the (normalized) angular form in $\mathbb{R}^{n_1}$ around the origin.

Consider now the commutative diagram

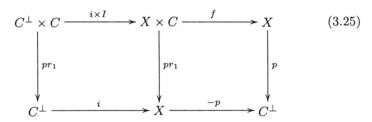

$$(3.25)$$

where $i =$ inclusion of $C^\perp$ to X, $f(x, y) = y - x$, $p =$ orthogonal projection of X to $C^\perp$.

Note that $f \circ (i \times I) : C^\perp \times C \to X$ is an isomorphism but does not preserve orientation; however it does preserve the orientation in the second factor C alone, i.e. in the fibers of the vertical maps.

Similarly, $-p \circ i : C^\perp \to C^\perp$ is multiplication by (-1) and so multiplies the orientation of $C^\perp$ by $(-1)^{n_1}$.

Now we recall that the angular form on $X \times X$ was $f^* d\theta$ and its restriction to $X \times C$ and image under pr_{1*} is $pr_{1*} f^* (d\theta)$.

Pulling back to $C^\perp$ by i^*, we get $i^*(pr_{1*} f^* d\theta) = pr_{1*}(i \times I)^* f^* d\theta$ (since the maps $i \times I$, f map fibers C homeomorphically, preserving orientation of C), which is also $= i^*(-p)^* p_*(d\theta) = $ (image) $p_*(d\theta)$ on $C^\perp = \mathbb{R}^{n_1}$, under the map $x \to -x$ of $\mathbb{R}^n$.

It remains to show that $p : \mathbb{R}^n \to \mathbb{R}^{n_1}$ (orthogonal projection) sends $d\theta$ to $p_*(d\theta) = d\theta_1$ on $C^\perp = \mathbb{R}^{n_1}$. We can then conclude that $i^*(-p)^*(d\theta_1) = i^* pr_{1*}(f^* d\theta) = (-1)^{n_1} d\theta_1$ on $C^\perp$, which is the restriction to $C^\perp$ of the angular form $d\theta_C$ around C.

This last step for $d\theta$ on $\mathbb{R}^n = \mathbb{R}^{n_1} \oplus \mathbb{R}^{n_2}$ comes from factoring $d\theta$ as follows, using $r^2 = x_1^2 + \ldots + x_n^2 = r_1^2 + r_2^2$:

$$d\theta = d\theta_1 \wedge \beta \wedge d\theta_2$$

$$\beta = B(n_1/2, n_2/2)^{-1}[(r_1^2/r^2)^{(n_1/2)-1}(r_2^2/r^2)^{(n_2/2)-1} d(r_2^2/r^2)]$$

where $B(n_1/2, n_2/2) = \int_0^1 (1 - \chi)^{n_1/2-1} \chi^{n_2/2-1} d\chi$ and $d\theta_i$ are the normalized angular forms of $\mathbb{R}^{n_i}$.

Then $p_*(d\theta) = $ integral of $d\theta$ with respect to the variables $x_{n_1+1}, \ldots, x_{n_1+n_2} = d\theta_1 \int \beta \int d\theta_2 = d\theta_1$.

This concludes the discussion of $\Gamma_C(X)$.

3.2   The main theorem

We can now state and prove our main result of this chapter, expressing a generalized linking number by iterated integrals. This can be done on a (compact, even dimensional) Riemannian manifold but if the manifold is complex and Kahler then the real number obtained is independent of the choice of Kahler metric and depends only on the complex structure.

Let Y be a smooth compact oriented Riemannian manifold of even dimension n, and let $C_1, \ldots, C_k$, $k \geq 3$, be smooth submanifolds of Y, or more generally let each C_i be a cycle which is an integer linear combination of smooth submanifolds. Let C_i have codimension p_i and assume

$$p_1 + \ldots + p_k = n + 1.$$

We will denote the support of C_i by C_i again and denote by $\bigcap$ the intersection of these supports. We make two assumptions:

   1.   The intersection of any $k - 1$ of the C_i is empty: $C_1 \cap \ldots \cap C_{i-1} \cap C_{i+1} \cap \ldots \cap C_k = \emptyset$ for $i = 1, \ldots, k$.

   2.   The intersection of any set of C_i is transverse (that is, if $C_i = \sum n_{ij} C_{ij}$ with C_{ij} smooth manifold and if $i_1, \ldots i_r$ are distinct indices then $C_{i_1 j_1}, \ldots, C_{i_r j_r}$ intersect transversely).

For simplicity we will write as if each C_i is an oriented manifold rather than a linear combination.

Clearly 1. is the main assumption and we will see that 2. is not a real restriction.

Let $X = Y^k = Y \times \ldots \times Y$ with product Riemannian metric.

Let the diagonal in Y^k, i.e. $\{(y, \ldots, y)\}$ be denoted Y, and let $C_1 \times \ldots \times C_k$ be the Cartesian product cycle in Y^k. Then the generalized linking number

$$(Y, C_1 \times \ldots \times C_k) \tag{3.26}$$

is defined. It will be expressed using the following iterated integrals: let ω_i be the harmonic (on Y) Poincare dual to the homology class of C_i. Then $\omega_1 \wedge \ldots \wedge \omega_{i-1}$ is a closed form Poincare dual to the (homology class of) the intersection cycle of $C_1, \ldots, C_{i-1}$, denoted $C_1 \bullet \ldots \bullet C_{i-1}$. By the assumption 1., this last cycle has disjoint support from the intersection cycle $C_{i+1} \bullet \ldots \bullet C_k$. To include the case $i = k$ in the notation, we define

$C_{k+1} = Y$ and write the two intersection cycles (with disjoint supports) as $C_1 \bullet \ldots \bullet C_{i-1}$ and $C_{i+1} \bullet \ldots \bullet C_{k+1}$ for $i = 1, \ldots, k$ (both being 0 for $i = 1$).

The closed form $\omega_1 \wedge \ldots \wedge \omega_{i-1}$ when restricted to the support of $C_{i+1} \bullet \ldots \bullet C_{k+1}$ is exact: there exist forms η_{i-1} on this support satisfying $d\eta_{i-1} = \omega_1 \wedge \ldots \wedge \omega_{i-1}$ there. We then write:

$$(\omega_1 \wedge \ldots \wedge \omega_{i-1}, \omega_i) = \eta_{i-1} \wedge \omega_i$$

$$(\omega_1 \wedge \ldots \wedge \omega_{i-1}, 1) = \eta_{i-1}$$

on this set $C_{i+1} \bigcap \ldots \bigcap C_{k+1}$. Since $\omega_1 \wedge \ldots \wedge \omega_k = 0$ we choose $\eta_k = 0$ ($\int_Y \eta_k = 0$ is all we need).

Write $q = \sum_{r<s} p_r p_s$ $(1 \le r < s \le k)$. We will need q only modulo 2: if j denotes the number of odd p_i, then $q \equiv j(j-1)/2 \bmod 2$ (j is odd since $\sum p_i = n+1$ is odd). The result is then ([Harris, 2002]):

Theorem 3.2   *With the above notation and for any choice of the forms η_{i-1} (with $\eta_k = 0$), we have equality between the generalized linking number on $X = Y^k$ on the left and iterated integrals on the right (recall $k \ge 3$):*

$$(Y, C_1 \times \ldots \times C_k) = (-1)^{q+1} \sum_{i=2}^{k} \int_{C_{i+1} \bullet \ldots \bullet C_{k+1}} [(\omega_1 \wedge \ldots \omega_{i-1}, \omega_i) - (\omega_1 \wedge \ldots \wedge \omega_i, 1)]$$

$$(3.27)$$

($q = \sum_{r<s} p_r p_s$ and the last term has $(\omega_1 \wedge \ldots \omega_k, 1) = 0$.

This expression (either side of the above equality) is unchanged if any C_i is replaced by a homologous cycle C_i' satisfying the same intersection conditions, i.e either side of the equation is a function of the homology classes of the C_i; if any two C_i, say C_r, C_s for $r \ne s$ are interchanged, the expression is multiplied by $(-1)^{p_r p_s}$.

If Y is complex and the metric is Kahler then the expression is unchanged if another Kahler metric on Y is used.

Proof: We denote points in $X \times X$ as (x, x'). $X = Y^k$ and we write $x = (y_1, \ldots, y_k)$, $x' = (y_1', \ldots, y_k')$. Since X has product metric, its Laplacian Δ satisfies $\Delta = \Delta_1 + \ldots + \Delta_k$, $\Delta_i =$ Laplacian on the ith factor Y, and the Δ_i commute as operators on X. Thus $\exp(-t\Delta)$ is the product of the $\exp(-t\Delta_i)$ and the kernel K_X for X is also a product.

$$K_X(x, x', t) = pr_1^* K_Y \wedge \ldots \wedge pr_k^* K_Y = K_Y(y_1, y_1', t) \wedge \ldots \wedge K_Y(y_k, y_k', t)$$

(we can also see this from the eigenform expansion). We abbreviate this as

$$K = K_1 \ldots K_k.$$

Similarly the harmonic part of K on $X \times X$ is $H_X = H = H_1 \ldots H_k$. Recall that $\Gamma_{X,t} = \Gamma(x, x', t)$ satisfied two conditions:
  (a) $d\Gamma_{X,t} = K_X - H_X$
  (b) $\Gamma_{X,t}$ is coexact (in image d^*).
If we change $\Gamma_{X,t}$ by adding an exact form

$$\Gamma'_{X,t} = \Gamma_{X,t} + dE,$$

then for cycles A, B on X with disjoint supports:

$$\int_{A \times B} \Gamma'_{X,t} = \int_{A \times B} \Gamma_{X,t} \to (A, B) \text{ as } t \to 0.$$

Thus to define the linking number (A, B) we can replace Γ by Γ' provided Γ' satisfies (a) and is orthogonal to harmonic forms on $X \times X$.
  For $X = Y^k$ we can choose

$$\Gamma'_{X,t} = \Gamma_{1,t}K_2 \ldots K_k + H_1\Gamma_{2,t}K_3 \ldots K_k + \ldots + H_1 \ldots H_{k-1}\Gamma_{k,t}$$

where $\Gamma_{i,t} = \Gamma(y_i, y'_i, t)$.
  The formula (a) for $d\Gamma'$ is easily checked since the K_i and H_i are even degree forms and d-closed. Orthogonality to harmonic forms on $X \times X$ is also obvious, since we can assume these to be products of harmonic forms on the factors $Y \times Y$ (corresponding to the eigenvalues 0 of Δ) and for each i the ith term of Γ' has a factor $\Gamma_{i,t}$ which is orthogonal to harmonic forms on the ith factor $Y \times Y$.
  We now have to integrate $\Gamma' = \Gamma'(y_1, \ldots, y_n, y'_1, \ldots, y'_n, t)$ over $(y'_1, \ldots, y'_n) \in C_1 \times \ldots C_k$ and $(y_1, \ldots, y_n) \in$ diagonal of Y^k.
  However the terms on the right hand side of the expression for Γ' list the y's and y''s in the order $y_1, y'_1, \ldots, y_n, y'_n$ and so we have to first move the y'_j to the right, leaving the y_j on the left. In the jth factor $(j = 1, \ldots, k)$ of the ith term we only need the differential forms of degree $n - p_j = $ dimension of C_j in the coordinates $(y'_{j_1}, \ldots, y'_{j_n})$ of y'_j, and so of degree p_j in y_j coordinates (for the factor H_j or K_j) or $p_j - 1$ in y_j (for the factor Γ_j). Thus in the ith term moving all dy' to the right introduces the sign

$$(-1)^{q+\pi_i}$$

where $q = \sum_{r<s} p_r p_s$ and $\pi_i = \sum_1^{i-1} p_j$ (since $n - p_j = p_j \bmod 2$).

We can now integrate the ith term over $C_1 \times \ldots \times C_k$. According to the Appendix on Orientations, this means that we integrate the jth factor over $y_j' \in C_j$ and multiply the resulting forms in the variables y_j. By the lemma in this appendix, this integration, denoted pr_{1*}, gives

$$pr_{1*}(H_j) = \omega_{C_j} = \omega_j \text{ (harmonic Poincare dual form to } C_j \text{ in } Y)$$

$$pr_{1*}K_j = K_{C_j}$$

$$pr_{1*}\Gamma_j = \Gamma_{C_j}$$

where $d\Gamma_{C_j} = K_{C_j} - \omega_{C_j}$ on Y
(Γ_{C_j} and K_{C_j} also depend on $t > 0$).
Thus denoting $pr_1 : X \times X \to X$ the first projection, we have:

$$pr_{1*}\Gamma'_{X,t} = \sum_{i=1}^{k}(-1)^{q+\pi_i}\omega_1(y_1)\wedge\ldots\wedge\omega_{i-1}(y_{i-1})\wedge\Gamma_{C_i,t}\wedge K_{C_{i+1},t}\wedge\ldots\wedge K_{C_k,t} \cdot$$

$$(3.28)$$

Next this has to be restricted to the diagonal Y: $y_1 = \ldots = y_k = y$ of Y^k, giving the wedge product on Y, and finally it has to be integrated over $y \in Y$, giving $(Y, C_1 \times \ldots C_k)$ when the limit $t \to 0$ is taken.
This last integral is then

$$\sum_{i=1}^{k}(-1)^{q+\pi_i}\int_Y \omega_1 \wedge \ldots \wedge \omega_{i-1} \wedge \Gamma_{C_i,t} \wedge \ldots \wedge K_{C_k,t} \qquad (3.29)$$

where $\omega_i = \omega_{C_i}$. In this sum the first term is $(-1)^q \int_Y \Gamma_{C_1,t}K_{C_2,t}\ldots K_{C_k,t}$ which by lemma 3.2 approaches 0 as $t \to 0$ (since $C_2 \cap \ldots \cap C_k$ is empty).
For $i \geq 2$ we write $C_1 \cap \ldots \cap C_{i-1} = C_{1\ldots i-1}$ and $w_{1\ldots i-1}$ for the Poincare dual harmonic form. Thus

$$w_{1\ldots i-1} = \omega_1 \wedge \ldots \wedge \omega_{i-1} + d\alpha_{i-1} \ (\alpha_{i-1} = \text{smooth form}) .$$

Also we have, as currents,

$$w_{1\ldots i-1} = \delta_{C_1\ldots i-1} - d\Gamma_{1\ldots i-1}$$

where $\Gamma_{1\ldots i-1}$ is smooth on the complement of $C_1 \cap \ldots \cap C_{i-1}$ and in particular is smooth on $C_{i+1} \cap \ldots \cap C_k \cap C_{k+1}$ which we define as X for $i = k$ (we take $C_{k+1} = X$ and for $i = k$, $\Gamma_{1\ldots k-1}$ is smooth on X).
Thus

$$\omega_1 \wedge \ldots \wedge \omega_{i-1} = \delta_{C_1\ldots i-1} - d(\Gamma_{1\ldots i-1} + \alpha_{i-1})$$

and the ith term in the integral is

$$(-1)^{q+\pi_i} \int_Y \delta_{C_{1...i-1}} \wedge \Gamma_{C_i,t} \wedge K_{C_{i+1},t} \wedge \ldots \wedge K_{C_k,t} -$$
$$(-1)^{q+\pi_i} \int_Y d(\Gamma_{1...i-1} + \alpha_{i-1}) \Gamma_{C_i,t} K_{C_{i+1},t} \ldots K_{C_k,t}.$$

The first of these two terms is, by definition of the Dirac current δ,

$$\pm \int_{C_1 \cap \ldots \cap C_{i-1}} \Gamma_{C_i,t} K_{C_{i+1},t} \ldots K_{C_k,t}$$

and since $C_1 \cap \ldots \cap C_{i-1} \cap C_{i+1} \cap \ldots \cap C_k$ is empty this term is 0 for $i = k$ and for $1 < i < k$ it is exactly the same as for $i = 1$ but with X replaced by $C_1 \cap \ldots \cap C_{i-1}$; thus as $t \to 0$ this term approaches 0. Thus we are left with the second of the above terms: here $d(\Gamma_{1...i-1} + \alpha_{i-1})$ is a current and the meaning of the integral (including the sign in front), since degree of $\Gamma_{1...i-1} + \alpha_{i-1}$ is $\pi_i - 1$, is:

$$(-1)^{q+1} \int_Y (\Gamma_{1...i-1} + \alpha_{i-1})(d\Gamma_{C_i,t}) \wedge K_{C_{i+1},t} \wedge K_{C_k,t}.$$

Write $\Gamma_{1...i-1} + \alpha_{i-1} = -\eta_{i-1}$, then this integral is a sum of two terms:

$$(-1)^q \int_Y \eta_{i-1} K_{C_i,t} \wedge \ldots \wedge K_{C_k,t} - (-1)^q \int_Y \eta_{i-1} \omega_i \wedge K_{C_{i+1},t} \wedge \ldots \wedge K_{C_k,t}.$$

In the first integral, η_{i-1} is an L^1 form, the angular form for $C_1 \cap \ldots \cap C_{i-1}$, and is smooth outside a neighborhood of this set (which does not meet $C_{i+1} \cap \ldots \cap C_k$). The integral over this neighborhood approaches 0 as $t \to 0$ since $K_{C_i,t} \wedge K_{C_{i+1},t} \wedge \ldots \wedge K_{C_k,t}$ approaches 0 rapidly here. $\eta_{1...i-1}$ being smooth over the complement of this neighborhood, which includes $C_i \cap \ldots \cap C_k$ on which $K_{C_i,t} \ldots K_{C_k,t}$ approaches $\delta_{C_1...C_k}$, this integral then approaches $(-1)^q \int_{C_i...C_k} \eta_{i-1}$.

In exactly the same way, but using disjointness of $C_1 \cap \ldots \cap C_{i-1}$ and $C_{i+1} \cap \ldots \cap C_k$, the second term approaches

$$-(-1)^q \int_{C_{i+1}...C_k} \eta_{i-1} \omega_i.$$

In each of these integrals, η_{i-1} is a smooth form on the domain of integration which is a manifold (or linear combination of manifolds) and ω_i, Poincare dual to C_i in X, is also Poincare dual to $C_i \ldots C_k$ in $C_{i+1} \ldots C_k$ (by our transversality hypothesis).

By Poincare duality in $C_{i+1} \bullet \ldots \bullet C_k$, if φ is a closed form on this manifold (or linear combination) then

$$\int_{C_i...C_k} \varphi - \int_{C_{i+1}...C_k} \varphi \wedge \omega_i = 0.$$

η_{i-1} is not a closed form: instead $d\eta_{i-1} = \omega_1 \wedge \ldots \wedge \omega_{i-1}$ on $C_{i+1} \ldots C_k$, and the above argument with closed forms φ says that

$$\int_{C_i \ldots C_k} \eta_{i-1} - \int_{C_{i+1} \ldots C_k} \eta_{i-1} \wedge \omega_i$$

depends only on $d\eta_{i-1}$, i.e. only on $\omega_1, \ldots \omega_{i-1}$ and $C_{i+1}, \ldots, C_k$. We can then write the above difference of two integrals as

$$\int_{C_i \ldots C_k} (\omega_1 \wedge \ldots \wedge \omega_{i-1}, 1) - \int_{C_{i+1} \ldots C_k} (\omega_1 \wedge \ldots \wedge \omega_{i-1}, \omega_i).$$

We recall that $\omega_1 \wedge \ldots \wedge \omega_k = 0$ (its degree is higher than $n = dimY$) and we choose $\eta_{1 \ldots k} = 0$. Now adding up the k integrals, the first being 0, we get the expression in the theorem (the last term in the sum being

$$(-1)^{q+1} \int_Y (\omega_1 \wedge \ldots \wedge \omega_{k-1}, \omega_k)) .$$

If $k = 3$ we find: for C_1, C_2, C_3 mutually disjoint,

$$(Y, C_1 \times C_2 \times C_3) = (-1)^{q+1} [\int_{C_3} (\omega_1, \omega_2) - (\omega_1 \wedge \omega_2, 1) + \int_Y (\omega_1 \wedge \omega_2, \omega_3)].$$

We note now that the iterated integral expression does not involve the cycle C_1, but only involves its harmonic Poincare dual ω_1. The iterated integrals are over $C_{i+1} \ldots C_{k+1}$ for $i \geq 2$.

We conclude that the linking number as well depends only on the homology class of C_1 (and on the metric).

Next we will prove a (super) symmetry of the linking number under interchanges of the C_i, which will also prove its dependence only on the homology classes of all the C_i. We will only give details for interchange of C_1, C_2, that is,

$$(Y, C_2 \times C_1 \times C_3 \times \ldots \times C_k) = (-1)^{p_1 p_2} (Y, C_1 \times C_2 \times \ldots \times C_k) \quad (3.30)$$

(recall $p_i = codim(C_i)$). To do this we just note that instead of carrying out the integration with the form $\Gamma'_{X,t}$ above, we can use

$$\Gamma''_{X,t} = K_1 \Gamma_{2,t} K_3 \ldots K_k + \Gamma_{1,t} H_2 K_3 \ldots K_k + \sum_{i=3}^k H_1 \ldots H_{i-1} \Gamma_{i,t} K_{i+1} \ldots K_k.$$

Then

$$(Y, C_2 \times C_1 \times C_3 \times \ldots \times C_k) = \lim_{t \to 0} \int_{Y \times (C_2 \times C_1 \times \ldots \times C_k)} \Gamma''_{X,t}$$

and the same calculation as before gives this as

$$(-1)^{p_1 p_2} (Y, C_1 \times C_2 \times \ldots \times C_k) .$$

Finally we assume X is complex with Kahler metric and examine the effect of changing to another Kahler metric with the same complex structure.

Denote with $'$ quantities depending on the second metric. Then ω_i is harmonic in the original metric and ω_i' is the corresponding harmonic form in the second metric. Then
$d\omega_i = 0 = d^c\omega_i$ and $d\omega_i' = 0 = d^c\omega_i'$ (d, d^c depend only on the complex structure) and $\omega_i' - \omega_i$ is d-exact. Since $\omega_i' - \omega_i$ is also d^c closed, the dd^c lemma says that there is a form λ_i on Y satisfying $\omega_i' - \omega_i = dd^c\lambda_i$ with $i = 1, \ldots, k$ (recall $dd^c = (i/2\pi)\partial\bar{\partial}$). Thus

$$\omega_1' \wedge \ldots \omega_i' = \omega_1 \wedge \ldots \wedge \omega_i + dd^c \sum_{j=1}^{i} \omega_1 \ldots \omega_{j-1}\lambda_j\omega_{j+1}' \ldots \omega_i'.$$

Having chosen any η_i on $C_{i+2} \bigcap \ldots \bigcap C_{k+1}$ satisfying $d\eta_i = \omega_1 \ldots \omega_i$ (and $\eta_k = 0$ or at least $\int_Y \eta_k = 0$) we can choose:

$$\eta_i' = \eta_i + d^c \sum_{j=1}^{i} \omega_1 \ldots \omega_{j-1}\lambda_j w_{j+1}' \ldots \omega_i'.$$

. On $C_{k+1} = X$ we get from $\eta_k = 0$ that

$$\eta_k' = d^c(\sum_{j=1}^{k} \omega_1 \ldots \omega_{j-1}\lambda_j\omega_{j+1}' \ldots \omega_k')$$

and so $\int_Y \eta_k' = 0$ since Y is a complex manifold.

For $i \leq k$, start with

$$\eta_{i-1}' = \eta_{i-1} + d^c \sum_{j=1}^{i-1} \omega_1 \ldots \omega_{j-1}\lambda_j w_{j+1}' \ldots \omega_{i-1}'$$

and multiply by $\omega_i' = \omega_i + dd^c\lambda_i$, getting

$$\eta_{i-1}'\omega_i' = \eta_{i-1}\omega_i + \eta_{i-1}dd^c\lambda_i + d^c \sum_{j=1}^{i-1}(\omega_1 \ldots \omega_{j-1}\lambda_j\omega_{j+1}' \ldots \omega_{i-1}')\omega_i'.$$

Subtracting this from the equation for η_i', we have

$$(\eta_i' - \eta_{i-1}'\omega_i') - (\eta_i - \eta_{i-1}\omega_i) = -\eta_{i-1}dd^c\lambda_i + d^c(\omega_1 \ldots \omega_{i-1}\lambda_i)$$

$$= -\eta_{i-1}dd^c\lambda_i + d^c[(d\eta_{i-1}\lambda_i].$$

Since $d^c d\eta_{i-1} = d^c\omega_1 \ldots \omega_{i-1} = 0$, this is

$$= -\eta_{i-1}dd^c\lambda_i + (-1)^{deg\eta_{i-1}+1}(d\eta_{i-1}) \wedge d^c\lambda_i$$

$$= d[(-1)^{deg\eta_{i-1}+1}\eta_{i-1} \wedge d^c\lambda_i]$$

and so the integral over a cycle, e.g. $C_{i+1} \ldots C_{k+1}$ is 0:

$$\int_{C_{i+1}...C_{k+1}} (\eta_i' - \eta_i' \omega_i') = \int_{C_{i+1}...C_{k+1}} (\eta_i - \eta_{i-1} \omega_i).$$

This concludes the proof of the theorem. $\square$

The following question now arises: is it possible to replace the disjointness hypothesis by a hypothesis on products of cohomology classes? Even the answer to the following basic question seems unknown: if X is a compact oriented manifold, simply connected, and Kahler, and α, β are cohomology classes with real coefficients whose cup product is 0, can their Poincare dual homology classes be represented by cycles with disjoint supports?

We will now state a "complex version" of this last theorem. X is again a compact complex Kahler manifold and $K(x, y, t)$, $H(x, y)$ are the heat kernel and the kernel for orthogonal projection to harmonic forms. $\gamma(x, y, t)$ is the coexact form on $X \times X$ (for each $t > 0$) satisfying:

γ is in image of $(dd^c)^*$, and

$$dd^c \gamma = K - H$$

(thus if X has complex dimension l then γ is of type $(l - 1, l - 1)$). For some formulas and details on γ, see [Harris, 1993] and [Harris, 2002] .

Let (A, B) be complex-analytic cycles on X which are disjoint and whose complex dimensions satisfy

$$dim_{\mathbb{C}} A + dim_{\mathbb{C}} B = dim_{\mathbb{C}} X - 1 .$$

We define the Archimedean Height Pairing $\langle A, B \rangle$ as

$$\langle A, B \rangle = \lim_{t \to 0} \int_{A \times B} \gamma(x, y, t). \tag{3.31}$$

If we integrate γ over A first we obtain a form $\gamma_A(y, t)$ and as $t \to 0$ one can show that $\gamma_A(y, t)$ approaches a current γ_A which is a smooth form outside of A and so can be integrated over B. Such currents are sometimes called "Green's forms" for A. However we need not elaborate on this and just use $\gamma(x, y, t)$.

Suppose now $X = Y^k = Y \times \ldots \times Y$ where Y is complex Kahler of complex dimension m and $C_1, \ldots, C_k$ ($k \geq 3$) are compact complex submanifolds of Y (or cycles which are linear combinations of submanifolds) and satisfy the following conditions as cycles in Y :

(a) $\sum_{i=1}^{k} codim(C_i) = (dimY)+1$ (all dimensions and codimensions are complex).

(b) Any $k-1$ of the C_i have empty intersection.

(c) All intersections are in "general position": if each C_i as cycle is $C_i = \sum a_{ij}C_{ij}$ (C_{ij} being submanifolds of Y), then for any distinct $i_1,\ldots,i_r$, $C_{i_1 j_1},\ldots,C_{i_r j_r}$ intersect in general position (and so the intersections are submanifolds).

Let $\alpha_1,\ldots,\alpha_k$ be the harmonic Poincare dual forms to the homology classes of the C_i.

Condition (b) implies that for $i = 2,\ldots,k$ there exist forms μ_{i-1} on $C_{i+1}\cap\ldots\cap C_k$ such that $\alpha_1 \wedge \ldots \wedge \alpha_{i-1} = dd^c\mu_{i-1}$ on this intersection. We set $\mu_k = 0$. (For instance, we could use the dd^c lemma, or else take $\mu_{i-1} = -\Gamma_Z$, where $Z = C_1 \bullet \ldots \bullet C_{i-1}$). For any choice of such μ_{i-1}, we write

$$(\alpha_1 \wedge \ldots \wedge \alpha_{i-1}, 1) = \mu_{i-1}$$

$$(\alpha_1 \wedge \ldots \wedge \alpha_{i-1}, \alpha_i) = \mu_{i-1} \wedge \alpha_i .$$

We then have

Theorem 3.3 *Let $X = Y^k$ with product metric (X, Y are compact Kahler). Let $C_1,\ldots,C_k$ be complex cycles on Y satisfying conditions a,b,c above, and let $\langle A, B\rangle$ denote the Archimedean pairing on X. Let $C_1 \times \ldots \times C_k$ be the product cycle on Y^k and Y be the diagonal cycle. Then*

$$\langle Y, C_1\times\ldots\times C_k\rangle = -\sum_{i=2}^{k} \int_{C_{i+1}\bullet\ldots\bullet C_{k+1}} (\alpha_1 \wedge\ldots\wedge\alpha_{i-1}, \alpha_i) - (\alpha_1 \wedge\ldots\wedge\alpha_i, 1)$$

(3.31)

(here $C_{k+1} = Y$). This real number (either side of the above equation) depends only on the homology classes of the C_i, and is unchanged under permutation of $C_1,\ldots,C_k$.

Proof: The proof is essentially the same as the proof of the previous theorem involving real cycles (but with fewer + or - signs as all dimensions are even). However the last statement of the previous theorem (independence of the metric) is not known to us to be true here since the proof does not carry over.

We remark that in the case where all the C_i are divisors we can choose as Green's currents $\gamma_{C_i} = \log\|\sigma_i\|^2$, σ_i being meromorphic sections of metrized analytic line bundles on Y (with the metric $\| \ \|$ normalized so that $dd^c\gamma_{C_i} = \delta_{C_i} - \alpha_{C_i}$, $\alpha_{C_i} = \alpha_i$ harmonic as before).

Comparing the calculation of the height pairing above with [Deligne 1987, Sec. 8] we see that he is studying the same height pairing for divisors C_i. Deligne does not assume the intersection condition (b) above; consequently our statement on homology classes does not hold in his situation.

Appendix: Orientations, Fiber Integration

It is necessary to orient a compact n-manifold in order to integrate an n-form over it. We make explicit here how we make various choices of orientations . All manifolds will be C^∞.

Let X be a compact oriented n-manifold and S a compact k-dimensional submanifold. An orientation of S is equivalent to an orientation of its normal bundle $N = N_S$ by the following convention: at $x \in S$ we choose an ordered basis of the tangent space $T_x(S)$ in the orientation class of S and follow it by an ordered basis of N_x. The resulting ordered basis of $T_x(X)$ is required to be in the orientation class of $T_x(X)$.

Our Poincare duality convention is then as follows: let ω_S be a closed form on X such that for every closed form α on X we have

$$\int_S \alpha = \int_X \alpha \wedge \omega_S .$$

Such ω_S exist and will be called "Poincare dual forms to S". For instance, ω could be a Thom form, Gaussian shaped in the normal direction to S and defining the normal orientation to S. Further, ω_S could depend on a real parameter $t > 0$ and as $t \to 0$, ω_S could approach the Dirac current δ_S of integration over S: this means that for all forms α (not necessarily closed) on X we have:

$$\int_X \alpha \wedge \omega_S \to \int_S \alpha \ \text{(written also as } \int_X \alpha \wedge \delta_S)$$

as $t \to 0$.

Next we consider two oriented submanifolds S_1, S_2 of X which intersect transversally so that their intersection T is a submanifold of dimension $dimT = dimS_1 + dimS_2 - n$. We orient T so that an oriented ordered basis for the normal bundle to T in X consists of an oriented ordered basis for the normal bundle to T in S_2 (which then also orients the normal bundle to S_1 in X) followed by a basis orienting the normal bundle to S_2 in X. Poincare dual forms $\omega_{S_1}, \omega_{S_2}$ to S_1, S_2 in X then determine a Poincare dual form ω_T:

$$\omega_T = \omega_{S_1} \wedge \omega_{S_2} \ .$$

Also, the restriction of ω_{S_1} to S_2 is a Poincare dual form to T in S_2.

We also have to orient Cartesian products of manifolds and of submanifolds.

If X_1, X_2 are oriented manifolds of dimensions n_1, n_2 we write $pr_i :$ $X_1 \times X_2 \to X_i$ for the projection, $i = 1, 2$, and for α_1, α_2 forms on X_1, X_2, we write $pr_1^* \alpha_1$ or $\alpha(x_1)$ and $pr_2^* \alpha_2$ or $\alpha_2(x_2)$ for the corresponding forms on $X_1 \times X_2$. If X_1, X_2 are oriented tangentially by top degree forms ω_1, ω_2, we orient $X_1 \times X_2$ by $pr_1^* \omega_1 \wedge pr_2^* \omega_2$. So

$$\int_{X_1 \times X_2} (pr_1^* \eta_1 \wedge pr_2^* \eta_2) = \int_{X_1} \eta_1 \int_{X_2} \eta_2 \ .$$

Let A_1, A_2 be oriented submanifolds of X_1, X_2 of dimensions a_1, a_2 and with Poincare dual closed forms $\omega_{A_1}, \omega_{A_2}$.

We would like to define the orientation and Poincare dual form $\omega_{A_1 \times A_2}$ of $A_1 \times A_2$ so that for any closed forms α_i on X_i, $deg \alpha_i = a_i$,

$$\int_{A_1 \times A_2} pr_1^* \alpha_1 \wedge pr_2^* \alpha_2 = \int_{A_1} \alpha_1 \int_{A_2} \alpha_2 = \int_{X_1 \times X_2} pr_1^* \alpha_1 \wedge pr_2^* \alpha_2 \wedge \omega_{A_1 \times A_2}.$$

For this we have to take

$$\omega_{A_1 \times A_2} = (-1)^{(n_1 - a_1)a_2} pr_1^* \omega_{A_1} \wedge pr_2^* \omega_{A_2} \ .$$

In particular, for $X_1 = A_1$ we get

$$\omega_{X_1 \times A_2} = pr_2^* \omega_{A_2} \ .$$

For $X_2 = A_2$ we get:

$$\omega_{A_1 \times X_2} = (-1)^{(n_1 - a_1)n_2} pr_1^* \omega_{A_1} \ .$$

Multiplying,

$$\omega_{X_1 \times A_2} \omega_{A_1 \times X_2} = \omega_{A_1 \times A_2}.$$

For n even $\omega_{A_1 \times A_2} = (-1)^{a_1 a_2} pr_1^* \omega_{A_1} \wedge pr_2^* \omega_{A_2}$.

As an example we consider in $X \times X$ the diagonal Δ (Δ should not be confused here with the Laplacian). Using local coordinates $x_1, \ldots, x_n$ in X, X is oriented by the n-form $dx_1 \wedge \ldots \wedge dx_n$. We take a copy Y of X with coordinates $y_1, \ldots, y_n$ ($\varphi^* y_i = x_i$ if φ is the identification map $\varphi : X \to Y$) and write the coordinates in $X \times X$ as $(x_1, \ldots, x_n, y_1, \ldots, y_n)$. We use the diagonal map $i : X \rightarrow X \times X$, $i(x_1, \ldots, x_n) = (x_1, \ldots, x_n, x_1, \ldots, x_n)$ and orient X and Δ so that i is orientation preserving as map from X to Δ. We orient X as above and orient the tangent bundle of X by the ordered basis $\frac{\partial}{\partial x_1}, \ldots, \frac{\partial}{\partial x_n}$, the tangent bundle of Δ by the basis $\frac{\partial}{\partial x_1} + \frac{\partial}{\partial y_1}, \ldots, \frac{\partial}{\partial x_n} + \frac{\partial}{\partial y_n}$, and the normal bundle of Δ in $X \times X$ by $-\frac{\partial}{\partial x_1} + \frac{\partial}{\partial y_1}, \ldots, -\frac{\partial}{\partial x_n} + \frac{\partial}{\partial y_n}$. Then $X \times X$ is oriented by $\frac{\partial}{\partial x_1}, \ldots, \frac{\partial}{\partial x_n}, \frac{\partial}{\partial y_1}, \ldots, \frac{\partial}{\partial y_n}$.

The projections pr_j, $j = 1, 2$, of $X \times X$ to X then induce orientation preserving diffeomorphisms of Δ with X.

Let now A_1, A_2 be oriented manifolds of dimensions k_1, k_2 and let $f_i : A_i \rightarrow X$ be differentiable maps. We assume that

$$f_1 \times f_2 : A_1 \times A_2 \rightarrow X \times X$$

is transverse to $\Delta \subset X \times X$. For example, this transversality holds if $A_1 = X$, $f_1 = Identity$. Then the fiber product of $f_1 \times f_2$ with the inclusion of Δ, in other words the inverse image of Δ in $A_1 \times A_2$, is a submanifold $A_1 \times_X A_2$ of $A_1 \times A_2$ with normal bundle in $A_1 \times A_2$ oriented by $(f_1 \times f_2)^*(\omega_\Delta)$ where ω_Δ orients the normal bundle of Δ in $X \times X$.

Assume now $A_1 = X$, $f_1 = $ Identity map I, write $F = I \times f : X \times A_2 \rightarrow X \times X$. Then $F^{-1}(\Delta) = $ graph of $f = \{(f(a_2), a_2) : a_2 \in A_2\}$, denote this by $gr(f)$. We claim:
$A_2 \rightarrow gr(f)$, induced by F, is an orientation preserving diffeomorphism (taking a to $(f(a), a)$. To see this, we take any top degree form β on A_2 that orients A_2 and show first that under $pr_2 : gr(f) \rightarrow A_2$, we have that $pr_2^*\beta$ is a top degree form on $gr(f)$ that gives an orientation of $gr(f)$. Namely, if the tangent space to A_2 at a point a_2 has oriented basis $\xi_1, \ldots \xi_k$, $k = dim(A_2)$, then at $(f(a_2), a_2)$, $gr(f)$ has tangent space basis $(f_*(\xi_1) \oplus \xi_1, \ldots, f_*(\xi_k) \oplus \xi_k)$. Assuming $\beta(\xi_1 \wedge \ldots \wedge \xi_k) = 1$, we find that

$$pr_2^*(\beta)[f_*(\xi_1) \oplus \xi_1, \ldots, f_*(\xi_k) \oplus \xi_k] = 1.$$

Next we recall that ω_Δ on $X \times X$ is locally

$$(-dx_1 + dy_1) \wedge \ldots \wedge (-dx_n + dy_n) \, .$$

We can then calculate that on $X \times A_2$, if $\beta = \beta_1 \wedge \ldots \wedge \beta_k$,

$$pr_2^*(\omega_\beta) \wedge [(Id \times f)^*(\omega_\Delta)] = (0 \oplus \beta_1) \wedge \ldots \wedge (0 \oplus \beta_k) \wedge ((-dx_1 + f^*dy_1) \wedge \\ \ldots \wedge (-dx_n + f^*dy_n) = f^*(\beta) \wedge (-1)^n dx_1 \wedge \ldots \wedge dx_n \, .$$

Now assume n is even, obtaining

$$dx_1 \wedge \ldots \wedge dx_n \wedge f^*(\beta)$$

which is the orientation of $X \times A_2$.

In conclusion, the orientation of $gr(f)$ by $(Id \times f)^* \omega_\Delta$ agrees, under the map $A_2 \to gr(f)$, with the orientation of A_2.

As our main calculation we will state the following lemma in which we use integration over the fiber pr_{1*}: this integration will be reviewed after the statement of the lemma. We denote by $f_* \delta_{A_2}$ the current on X image of δ_{A_2} on A_2.

Lemma. *As before denote by F the map $Id \times f : X \times A_2 \to X \times X$ and by $gr(f)$ the inverse image $F^{-1}(\Delta) \subset X \times A_2$. Let $pr_1 : X \times A_2 \to X$ be the projection and $pr_{1*} : A^p(X \times A_2) \to A^{p-k}(X)$ $(k = dim(A_2))$ (degree p currents on $X \times A_2$) $\to$ (degree $p - k$ currents on X) the integration over the fiber map. If ω_Δ is any closed form on $X \times X$, Poincare dual to Δ, then $pr_{1*}(F^* \omega_\Delta)$ is a closed form ω_{A_2} on X, Poincare dual to the cycle $f_*(A_2)$ in X. If $k_{t,\Delta}$ for each $t > 0$ is a Gaussian-shaped Thom form on $X \times X$ which as $t \to 0$ approcahes the current δ_Δ then $F^*(k_{t,\Delta})$ is a similarly shaped form on $X \times A_2$ approaching the current $\delta gr(f)$ as $t \to 0$, and $pr_{1*} F^*(k_{t,\Delta}) = k_{t,A_2}$ approaches $f_*(\delta_{A_2})$ on X as $t \to 0$. Finally, if $\Gamma(x,y,t)$ is for $t > 0$ a smooth form on $X \times X$ satisfying $d\Gamma = k_{t,\Delta} - \omega_\Delta$ then $\Gamma_{t,A_2} = pr_{1*} F^*(\Gamma)$ is a C^∞ form on X satisfying $d\Gamma_{t,A_2} = k_{t,A_2} - \omega_{A_2}$.*

Proof: 1.To prove the statement about ω_Δ, let α be any closed form on X, then

$$\int_X \alpha \wedge pr_{1*} F^*(\omega_\Delta) = \int_{X \times A_2} pr_1^* \alpha \wedge F^*(\omega_\Delta) = \int_{gr(f)} pr_1^* \alpha$$

(since $F^*(\omega_\Delta)$ is a Poincare dual form to $gr(f)$). Now using the orientation preserving diffeomorphism $f \times Id : A_2 \to gr(f)$ and the map $pr_1 \circ (f \times Id) = f : A_2 \to X$ we get:

$$\int_{gr(f)} pr_1^* \alpha = \int_{A_2} (f \times Id)^* pr_1^* \alpha = \int_{A_2} f^* \alpha = f_*(\delta_{A_2})(\alpha).$$

2. For any form α on X,

$$\int_X \alpha \wedge pr_{1*}(F^* k_{t,\Delta}) = \int_{X \times A_2} pr_1^* \alpha \wedge (F^* k_{t,\Delta}).$$

Write $F^*(k_{t,\Delta}) = k_{t,gr(f)}$: since F is transverse to Δ, $k_{t,gr(f)}$ is a Gaussian shaped form on $X \times A_2$ peaking on $gr(f)$. Thus the last integral above approaches $\int_{gr(f)} pr_1^* \alpha$ as $t \to 0$, and as in 1., this equals $\int_{A_2} f^*(\alpha) = f_*(\delta_{A_2})\alpha$.

3. Starting with $d\Gamma(x,y,t) = k_{t,\Delta} - \omega_\Delta$ on $X \times X$ we apply F^* and then pr_{1*}, both of which commute with d, obtaining the statement for Γ_t, k_{t,A_2}, ω_{A_2} on X. $\square$

Integration over the fiber

Denote by X_1, X_2 compact oriented manifolds of dimensions n_1, n_2 and by $pr_i : X_1 \times X_2 \to X_i$ the projection. Denote by A^* the differential forms on these manifolds. We define:

$$pr_{1*} : A^*(X_1 \times X_2) \to A^*(X_1)$$

in such a way that for $\alpha_i \in A^*(X_i)$,

$$pr_{1*}(pr_1^*\alpha_1 \wedge pr_2^*\alpha_2) = \alpha_1 (\textstyle\int_{X_2} \alpha_2)$$

[similarly $pr_{2*}(pr_1^*\alpha_1 \wedge pr_2^*\alpha_2) = (\int_{X_1} \alpha_1)\alpha_2$].
Then for $\varphi \in A^*(X_1)$, $\psi \in A^*(X_1 \times X_2)$

$$pr_{1*}[(pr_1^*\varphi) \wedge \psi] = \varphi \wedge pr_{1*}(\psi) \in A^*(X_1)$$

$$\textstyle\int_{X_1 \times X_2}[(pr_1^*\varphi) \wedge \psi] = \int_{X_1}[\varphi \wedge pr_{1*}(\psi)].$$

If X_1, X_2 have no boundary then

$$d \circ pr_{1*} = pr_{1*} \circ d$$

$$d \circ pr_{2*} = (-1)^{n_1} pr_{2*} \circ d \ (n_1 = dim X_1).$$

We extend pr_i^* to a map from currents T_i on X_i to currents $pr_i^*(T_i)$ on $X_1 \times X_2$ in such a way that if T_i is given by a form τ_i on X_i (thus $T_i(\alpha_i) = \int_{X_i} \alpha_i \wedge \tau_i$) then pr_i^* is given by the form $pr_i^*(\tau_i)$.

Thus we define, for $\alpha_i \in A^*(X_i)$,

$$pr_2^*(T_2)(pr_1^*\alpha_1 \wedge pr_2^*\alpha_2) = (\textstyle\int_{X_1} \alpha_1)T_2(\alpha_2) = T_2(pr_{2*}(pr_1^*\alpha_1 \wedge pr_2^*\alpha_2))$$

[in particular, $pr_2^*(\delta_{A_2}) = \delta_{X_1 \times A_2}$] and

$$pr_1^*(T_1)[pr_1^*\alpha_1 \wedge pr_2^*\alpha_2] = (-1)^{(deg T_1)(deg \alpha_2)} T_1(\alpha_1) \textstyle\int_{X_2} \alpha_2 \ \text{(assuming } deg \alpha_2 = n_2).$$

[In particular $pr_1^*\delta_{A_1} = (-1)^{(n_1-a_1)n_2}\delta_{A_1 \times X_2}$.]
We thus have
$pr_i^*(T_i) = T_i \circ pr_{i*} : A^*(X_1 \times X_2) \to \mathbb{R}$ for all n_1 and even n_2.
For a general map $f : X \to Y$ we define, for a current T on X, form α on Y,

$$f_*(T)(\alpha) = T(f^*\alpha).$$

Integration over the fiber

Bibliography

NOTES

We would like to point out in particular the following references: 1. K.T. Chen's "Collected Papers" and the summary in it of his life and work by R. Hain and P. Tondeur. 2. The survey article by R. Hain, "Iterated Integrals and Algebraic Cyles: Examples and Prospects" in the volume "Contemporary Trends in Algebraic Geometry and Algebraic Topology" edited by S.S. Chern, L. Fu and R. Hain (*Nankai Tracts in Mathematics* vol. 5, World Scientific, 2002), and many other valuable papers by R. Hain. 3. The book "Heat Kernels and Dirac Operators" by N.Berline, E.Getzler, and M. Vergne. 4. The article "Groupes Fondamentaux Motiviques de Tate Mixte" by P. Deligne and A.B. Goncharov, giving some different viewpoints and directions from those in this book. 5. The survey article "The Ubiquitous Heat Kernel" by Jorgenson and Lang. 6. Many articles by R. Harvey and B. Lawson on currents, differential characters and singularities.

THE REFERENCES

Berline, N., Getzler, E., Vergne, M. (1992). Heat Kernels and Dirac Operators. Springer, Berlin.

Bloch, S. (1984). Algebraic Cycles and Values of L-functions. J.Reine.Angew.Math. 350, pp. 899–912.

Chen, K.T. (2000). Collected Papers of K.T.Chen. P.Tondeur, Editor. Birkhauser, Boston.

Deligne, P. (1987). Le Determinant de la Cohomologie. Contemporary Mathematics 67, pp. 93–177.

Deligne, P. et Goncharov, A.B. (2003). Groupes Fondamentaux Motiviques de Tate Mixte. ArXiv:Math.NT/0302267.

Griffiths, P. and Harris, J. (1978). Principles of Algebraic Geometry. Wiley.

Hain, R. (2002). Iterated Integrals and Algebraic Cycles: Examples and Prospects. In: Contemporary Trends in Algebraic Geometry and Algebraic Topology, Editors S.S. Chern, L. Fu, R. Hain. World Scientific.

Harris, B. (1983). Harmonic Volumes. Acta Mathematica 150, pp. 91–123.

Harris, B. (1983). Homological vs Algebraic Equivalence in a Jacobian. Proc.Nat.Acad. of Sciences USA 80, pp. 1157-1158.

Harris, B. (1989). Differential Characters and the Abel-Jacobi map, pp. 69–86. Algebraic K-Theory: Connections with Geometry and Topology, Ed. by J.Jardine and V.Snaith. Kluwer.

Harris, B. (1990). Iterated Integrals and Epstein Zeta Function with Harmonic Rational Fucntion Coefficients. Illinois J.Math 34. pp. 325–336.

Harris, B. (1993). Cycle Pairings and the Heat Equation. Topology 32. pp. 225–238.

Harris, B. (2002). Chen's Iterated Integrals and Algebraic Cycles. pp. 119–134. In: Contemporary Trends in Algebraic Geometry and Algebraic Topology, Ed. by S.S. Chern, L. Fu, and R.Hain. World Scientific.

Hein, G. (2001). Computing Green Currents via the Heat Kernel. J.Reine und Angew.Math. 540, pp. 87–104.

Jorgenson, J. and Lang, S. (2001). The Ubiquitous Heat Kernel, pp. 655–682. In: Mathematics Unlimited: 2001 and Beyond, Ed. by B. Engquist, W. Schmid. Springer.

Kodaira, K. and Morrow, J. (1971). Complex Manifolds. Holt, Rinehart and Winston.

Lazard, M. (1954). Sur les Groupes Nilpotents et les Anneaux de Lie. Ann.Ecole Norm. Super. 71, pp. 101–190.

Quillen, D.G. (1968). On the Associated Graded Ring of a Group Ring. Journal of Algebra 10, pp. 411–418.

Schiffer, M. and Spencer, D.C. (1954). Functionals of Finite Riemann Surfaces. Princeton University Press.

Weil, A. (1962). Foundations of Algebraic Geometry (2nd edition), pp. 331. American Mathematical Society.

Weil, A. (1979). Collected Papers. Paper 1952e and Comments. Springer.

List of Notations

L free Lie algebra   10

H Hopf algebra   10, 11

$P(H)$ primitive elements of a Hopf algebra H 12

$L^\wedge, \mathfrak{g}^\wedge$ completions of the graded Lie algebras $L, \mathfrak{g}$
(formal infinite series)   12

$L_{\geq n}, \mathfrak{g}_{\geq n}$ all elements of degree $\geq n$, $\mathfrak{g}^\wedge = \varprojlim(\mathfrak{g}/\mathfrak{g}_{\geq n})$ 12, 13

$\Delta^\wedge$ extension of $\Delta : U(\mathfrak{g}) \to U(\mathfrak{g}) \otimes U(\mathfrak{g})$ to a homomorphism
$U(\mathfrak{g})^\wedge \to (U(\mathfrak{g}) \otimes U(\mathfrak{g}))^\wedge$ 13

$\mathfrak{g}_{(n)}$ $\mathfrak{g}/\mathfrak{g}_{\geq n}$ (Lie algebra)   15

$G_{(n)}$ Lie group of $(\mathfrak{g}/\mathfrak{g}_{\geq n})$ 15

$h_{(n)}$ group homomorphism $\pi_1 \to G_{(n)}$ 15

$I\pi$ same as $\overline{\mathbb{R}\pi}$ 16

$\mathcal{H}^i, C^1, E^2$ special subspaces of the differential forms $A^i(X)$ 16, 34

$\mathcal{H}^p$ harmonic p-forms   17

Q Lie algebra homomorphism induced by q 20

$\pi^{(2)}, \pi^{(n)}$ $(\pi, \pi), (\pi^{(n-1)}, \pi)$ terms of lower central series   20

$B(\pi), K(\pi, 1)$ classifying space of a discrete group π 22

π_i i-th homotopy group   22

$H_2^{sph}(X)$ image of $\pi_2(X) \to H_2(X)$, spherical homology classes   23

π_1^{ab} $\pi_1/(\pi_1, \pi_1)$

Λ^i i-th exterior power   25

ϕ, ψ Lie algebra homomorphisms   25, 26

$\mathfrak{g}^{[n]}$ $[[\mathfrak{g}_1, \mathfrak{g}_1], \cdots, \mathfrak{g}_1](n$ factors$)$ 27

$G^{(n)}$ $\exp \mathfrak{g}^{[n]}$

g Riemannian metric   29

$d\,vol$ volume element   29

$\star$ Hodge star operator $A^p(X) \to A^{n-p}(X)$ 29

d^* adjoint of d 30

Δ Laplace operator: $dd^* + d^*d$ (sometimes denotes diagonal).   30

$P(\gamma)$ orthogonal projection of γ in $A^p(X)$ to $\mathcal{H}^p$ 30

J almost complex structure operator, $T_x(X) \to T_x(X)$ 31

$\omega(v, w)$ Kähler 2-form   32

d^c operator $\frac{1}{4\pi}(J^{-1} \circ d \circ J)$ on forms, also $= i(\bar{\partial} - \partial)/4\pi$ 33

$(d^c)^*$ adjoint of d^c, also $= J^{-1}d^*J/4\pi$ 33

$\bar{\theta}$ simplified version of θ 35, 36

$\mathcal{H}_{\mathbb{Z}}^1$ harmonic 1-forms with periods in $\mathbb{Z}$ 36

$(\mathcal{H}_{\mathbb{Z}}^1 \otimes \mathcal{H}_{\mathbb{Z}}^1)'$ kernel of intersection number pairing   37

Index

L^1 current (and L^1 form), 72
d, d^c lemma, 34

Abel-Jacobi map, 42
algebraic equivalence of cycles, 52
angular current, 82
angular form, 76, 80–83, 88
Archimedean height pairing, 91
asymptotic expansion, 78, 82

Chen's connection, 7, 16
Chen's Lie algebra, 14
coexact forms, 16, 34
completion (inverse limit), 12, 14
current, 67
cycle, 78, 84

descending central series, 19
diagonal homomorphism, 9, 11
Dirac current, 67

Euclidean heat kernel, 76

Fermat quartic curve, 54
first Chern form, 69
free Lie algebra, 9

Green's currents, 93
Green's forms, 91
group homology, 22
group-like element, 11

harmonic forms, 30
harmonic volume, 42
heat equation, 75
heat kernel, 74
heat operator, 73, 74
Hodge $*$-operator, 29
Hodge theory, 28
Hopf algebra, 10, 11
hyperelliptic Riemann surfaces, 50

integration over the fiber, 99
intermediate Jacobian, 46
iterated integral, 1

Jacobian manifold, 45
Jacobian variety, 45

Kahler manifold, 32
Kahler metric, 34

Laplace operator, 30
lattice (discrete cocompact)
    subgroup, 26
linking number, 73, 76, 84–86, 89
lower central series, 7

Maurer-Cartan 1-form, 3
modular curve, 54

orientations, 95

period matrix, 60

107